Periodic Table of the Elements with the Gmelin System Numbers

Each cell shows: atomic number, element symbol, and Gmelin System Number.

1 H 2																	2 He 1
3 Li 20	4 Be 26											5 B 13	6 C 14	7 N 4	8 O 3	9 F 5	10 Ne 1
11 Na 21	12 Mg 27											13 Al 35	14 Si 15	15 P 16	16 S 9	17 Cl 6	18 Ar 2
19 K 22 *	20 Ca 28	21 Sc 39	22 Ti 41	23 V 48	24 Cr 52	25 Mn 56	26 Fe 59	27 Co 58	28 Ni 57	29 Cu 60	30 Zn 32	31 Ga 36	32 Ge 45	33 As 17	34 Se 10	35 Br 7	36 Kr 1
37 Rb 24	38 Sr 29	39 Y 39	40 Zr 42	41 Nb 49	42 Mo 53	43 Tc 69	44 Ru 63	45 Rh 64	46 Pd 65	47 Ag 61	48 Cd 33	49 In 37	50 Sn 46	51 Sb 18	52 Te 11	53 I 8	54 Xe 1
55 Cs 25	56 Ba 30	57** La 39	72 Hf 43	73 Ta 50	74 W 54	75 Re 70	76 Os 66	77 Ir 67	78 Pt 68	79 Au 62	80 Hg 34	81 Tl 38	82 Pb 47	83 Bi 19	84 Po 12	85 At 8a	86 Rn 1
87 Fr 25a	88 Ra 31	89*** Ac 40	104 71	105 71													

**Lanthanides 39

58 Ce	59 Pr	60 Nd	61 Pm	62 Sm	63 Eu	64 Gd	65 Tb	66 Dy	67 Ho	68 Er	69 Tm	70 Yb	71 Lu

***Actinides

90 Th 44	91 Pa 51	92 U 55	93 Np 71	94 Pu 71	95 Am 71	96 Cm 71	97 Bk 71	98 Cf 71	99 Es 71	100 Fm 71	101 Md 71	102 No 71	103 Lr 71

* NH_4 23

Gmelin Handbook of Inorganic and Organometallic Chemistry

8th Edition

Gmelin Handbook of Inorganic and Organometallic Chemistry

8th Edition

Gmelin Handbuch der Anorganischen Chemie

Achte, völlig neu bearbeitete Auflage

PREPARED
AND ISSUED BY

Gmelin-Institut für Anorganische Chemie
der Max-Planck-Gesellschaft
zur Förderung der Wissenschaften

Director: Ekkehard Fluck

FOUNDED BY   Leopold Gmelin

8TH EDITION   8th Edition begun under the auspices of the
Deutsche Chemische Gesellschaft by R. J. Meyer

CONTINUED BY   E. H. E. Pietsch and A. Kotowski, and by
Margot Becke-Goehring

Springer-Verlag Berlin Heidelberg GmbH 1995

Organometallic Compounds in the Gmelin Handbook

The following listing indicates in which volumes these compounds are described:

Ag   Silber B 5 (1975)

Au   Organogold Compounds (1980)

Be   Organoberyllium Compounds 1 (1987)

Bi   Bismut-Organische Verbindungen (1977)

Co   Kobalt-Organische Verbindungen 1, 2 (1973); Kobalt Erg.-Bd. A (1961), B 1 (1963), B 2 (1964)

Cr   Chrom-Organische Verbindungen (1971)

Cu   Organocopper Compounds 1 (1985), 2 (1983), 3 (1986), 4 (1987), Index (1987)

Fe   Eisen-Organische Verbindungen A 1 (1974), A 2 (1977), A 3 (1978), A 4 (1980), A 5 (1981), A 6 (1977), A 7 (1980), Organoiron Compounds A 8 (1986), A 9 (1989), A 10 (1991), A 11 (1995), Eisen-Organische Verbindungen B 1 (1976), Organoiron Compounds B 2 (1978), Eisen-Organische Verbindungen B 3 (1979), B 4, B 5 (1978), Organoiron Compounds B 6, B 7 (1981), B 8, B 9 (1985), B 10 (1986), B 11 (1983), B 12 (1984), B 13 (1988), B 14, B 15 (1989), B 16a, B 16b, B 17 (1990), B 18 (1991), B 19 (1992), Eisen-Organische Verbindungen C 1, C 2 (1979), Organoiron Compounds C 3 (1980), C 4, C 5 (1981), C 6a (1991), C 6b (1992), C 7 (1985); Eisen B (1929–1932)

Ga   Organogallium Compounds 1 (1986)

Ge   Organogermanium Compounds 1 (1988), 2 (1989), 3 (1990), 4 (1994), 5 (1993)

Hf   Organohafnium Compounds (1973)

In   Organoindium Compounds 1 (1991)

Mo   Organomolybdenum Compounds 5 (1992), 6 (1990), 7 (1991), 8 (1992), 9 (1993), 10 (1995), 12 (1994)

Nb   Niob B 4 (1973)

Ni   Nickel-Organische Verbindungen 1 (1975), 2 (1974), Register (1975), Organonickel Compounds Suppl. Vol. 1 (1993), 2 (1994); Nickel B 3 (1966), C 1 (1968), C 2 (1969)

Np, Pu   Transurane C (partly in English; 1972)

Os   Organoosmium Compounds A 1 (1992), A 2 (1993), B 3 (1994), B 4a (1995) **present volume**, B 5 (1994), B 6 (1993), B 8 (1995), B 9 (1995)

Pb   Organolead Compounds 1 (1987), 2 (1990), 3 (1992)

Po   Polonium Main Volume (1941, reprint 1969), Suppl. Vol. 1 (1990)

Pt   Platin C (1939) and D (1957)

Re   Organorhenium Compounds 1, 2 (1989), 3 (1992), 5 (1994)

Ru   Ruthenium Erg.-Bd. (1970)

Sb   Organoantimony Compounds 1, 2 (1981), 3 (1982), 4 (1986), 5 (1990)

Sc, Y,   Rare Earth Elements D 6 (1983)
La to Lu

Sn   Zinn-Organische Verbindungen 1 (1975) to 6 (1979), Organotin Compounds 7 (1980) to 23 (1995)

Ta   Tantal B 2 (1971)

Ti   Titan-Organische Verbindungen 1 (1977), 2 (1980), Organotitanium Compounds 3 (1984), 4 (1984), 5 (1990)

U   Uranium Suppl. Vol. E 2 (1980)

V   Vanadium-Organische Verbindungen (1971); Vanadium B (1967)

Zr   Organozirconium Compounds (1973)

Gmelin Handbook of Inorganic and Organometallic Chemistry

8th Edition

Os
Organoosmium
Compounds

Part B 4 a

With 59 illustrations

AUTHORS   Karin Greiner
Petra Olms–Keller (Göttingen)

EDITOR   Cornelia Weber

FORMULA INDEX   Bernd Kalbskopf
Hans–Jürgen Richter–Ditten

CHIEF EDITOR   Johannes Füssel

Springer-Verlag Berlin Heidelberg GmbH 1995

LITERATURE CLOSING DATE: MID – 1994
IN SOME CASES MORE RECENT DATA HAVE BEEN CONSIDERED

Library of Congress Catalog Card Number: Agr 25-1383

ISBN 978-3-662-07541-8   ISBN 978-3-662-07539-5 (eBook)
DOI 10.1007/978-3-662-07539-5

© by Springer-Verlag Berlin Heidelberg 1995
Originally published by Springer-Verlag Berlin Heidelberg New York London Paris Tokyo HongKong Barcelona Budapest Milan in 1995
Softcover reprint of the hardcover 8th edition 1995

Preface

The **Gmelin series "Organometallic Compounds"** comprises compounds containing at least one carbon–to–metal bond (except cyano compounds, which are considered inorganic). It includes all information in scientific journals, but patents, conference reports, and dissertations generally were not reviewed. The volumes published so far are listed on p. VI.

Organometallic compounds are classified according to their nuclearity and the bonding mode of the organic ligands ^nL. Nuclearity means the number of atoms of the title metal in the formula unit disregarding any additional metals that may be present. The term ^nL designates a ligand bonded by n carbon atoms to one or different atoms of the title metal. As usual, σ–bonded ^1L ligands are designated by R.

Inorganic ligands (i.e., ligands bonded exclusively by elements other than carbon) are generally designated by D or X. D means donor ligands such as pyridine or phosphanes; m–electron donors are specified by ^mD. X is reserved for negatively charged ligands or other one–electron donors such as halogens or SnR_3; bridging X ligands may donate one (μ–H), three (μ–Cl, μ–OR), or five (μ_3–I) electrons. Terms such as $^1L-^2D$, ^2D-X, or $^2D-^2D-^2D$ may be used for multidentate ligands. Heterometals are often designated by M, and bridging elements, bridging groups, or nonmetallic cluster constituents by E. The symbols η and μ follow the IUPAC nomenclature.

The **series "Organoosmium Compounds"** is divided into Series A dealing with mononuclear compounds and Series B treating di– to polynuclear compounds. Volume B 1 will be devoted to Os_2 compounds, volumes B 2 to B 7 to Os_3 compounds, volume B 8 to Os_4 compounds, and volume B 9 to compounds with higher nuclearity. Up to now, volumes A 1 (1992), A 2 (1993), B 3 (1994), B 5 (1994), B 6 (1993), and B 9 (1995) have been published.

The **present volume**, "Organoosmium Compounds" B 4a, systematically covers the literature through mid–1994, including some later references. In the first portion, the description of Os_3 carbonyls with additional N–bonded ligands, started in "Organoosmium Compounds" B 3 (1994), has been brought to a close. This part deals with all Os_3 carbonyls containing bridging N–bonded ligands. A second part is devoted to Os_3 carbonyls with additional P–bonded ligands. It contains all compounds with up to 8 CO groups; compounds with more than 8 CO groups will be treated in "Organoosmium Compounds" B 4b (in preparation).

The structure of the compounds in the present volume is mostly well established. As far as possible, the symbols μ and η are used in the linearized formulas. Hydride ligands generally bridge metal–metal edges in a $μ_2$ fashion. Since, however, terminal hydrides are also known, hydrides are only specified as μ–H in the formulas if these structures were either confirmed or are strongly suggested.

The volume is concluded with an empirical formula index, a ligand formula index, and a transition metal cross reference.

Many of the data, particularly in tables, are given in an abbreviated form and without units. For explanations, abbreviations, and units used throughout this volume, see p. X. Additional remarks are made if necessary at the table headings.

Frankfurt am Main, October 1995   Johannes Füssel

Explanations, Abbreviations, and Units

Most compounds in this volume are presented in tables. For the sake of conciseness, some abbreviations are used and some dimensions are omitted in the tables. This necessitates the following clarification.

Formulas. For conciseness or for better comparison of associated compounds, labeling can deviate from IUPAC numbering. **Geometric isomers** are designated according to the IUPAC rules. Structural labels are missing when authors fail to report structural details.

Abbreviations used with **temperatures** are m.p. for melting point, b.p. for boiling point, subl. for sublimation temperature, dec. for decomposition.

Solvents or the physical state and temperatures are given in parentheses directly after the spectral symbol if reported; none is given if room temperature applies.

Preparations and reactions were generally carried out under an inert-gas atmosphere. Abbreviations used with workup methods are TLC (thin-layer chromatography), GC (gas chromatography), PLC (preparative layer chromatography), and HPLC (high-pressure liquid chromatography). They were performed with silica gel, if not otherwise stated.

Nuclear magnetic resonance is abbreviated as NMR, noise decoupling (if mentioned by the authors) is indicated by braces { }. Chemical shifts are given as δ values in ppm with the positive sign for downfield shifts, differing signs given by authors are corrected if possible from the given information. Reference substances are $Si(CH_3)_4$ for ^1H and ^{13}C NMR, $BF_3 \cdot O(C_2H_5)_2$ for ^{11}B NMR, $CFCl_3$ for ^{19}F NMR, and H_3PO_4 for ^{31}P NMR, if not stated otherwise.

Multiplicities of the signals are abbreviated as s, d, t, q (singlet to quartet), quint, sext, sept (quintet to septet), and m (multiplet); terms like dd (double doublet), s's (singlets), and pt or vt (pseudo or virtual triplet) are also used. Coupling constants ^nJ in Hz are given as J(A,B) or as J(1,3) referring to labeled structural formulas, n is the number of bonds between the coupled nuclei.

Optical spectra are labeled as IR (infrared), Raman, and UV/vis (electronic spectrum comprising the ultraviolet and visible region). IR and Raman bands are given in cm^{-1}; in the band assignments the symbols ν, δ, ϱ, τ, χ, and γ are used for stretching, deformation, rocking, twisting, wagging, and out-of-plane bending vibrations, but ν is omitted if not necessary. The indexes s and as mean symmetrical and asymmetrical. The UV absorption maxima, λ_{max}, are given in nm, followed by the extinction coefficient ε ($L \cdot cm^{-1} \cdot mol^{-1}$) or log ε in parentheses; sh means shoulder, br means broad.

Electron spin resonance is abbreviated by ESR. Hyperfine interactions are characterized by $a_{nucleus}$.

Mass spectral data are given as the most important ions or the m/e values followed by the relative intensities in parentheses. Abbreviations for ionization methods are EI (electron impact), FAB (fast atom bombardment), FD (field desorption), and CI (chemical ionization). The molecular ion is abbreviated with $[M]^+$.

Further abbreviations:

conc.	concentrated	THF	tetrahydrofuran
D_c	calculated density	$i-C_3H_7$	isopropyl $CH(CH_3)_2$
D_m	experimental density	$i-C_4H_9$	isobutyl $CH_2CH(CH_3)_2$
vs.	versus	$t-C_4H_9$	tert-butyl $C(CH_3)_3$
$Cr(acac)_3$		tris-(2,4-pentandionato)-chrom	
18-crown-6/tetraglyme		1,4,7,10,13,16-hexaoxacyclooctadecane/$CH_3O(CH_2)_4OCH_3$	

Table of Contents

Organoosmium Compounds B 4a

3 Trinuclear Compounds (continued)

3.1 Compounds with Ligands Bonded by One Carbon Atom (continued)

3.1.1 Compounds with Carbonyl Ligands (continued)

3.1.1.7 Compounds with Additional N-Bonded Ligands (continued)

$Os_3(CO)_{10}(NCCH_3)_2$, $Os_3(CO)_{11}NCCH_3$, and other triosmium carbonyl compounds with additional terminal N-bonded ligands have been treated in "Organoosmium Compounds" B 3, 1994, Sections 3.1.1.7.1.1 to 3.1.1.7.1.3, pp. 218/60.

3.1.1.7.2 Compounds with η^1-Bonded Bridging N Ligands

3.1.1.7.2.1 Compounds with μ_2-η^1-Bonded NHR Ligands

The compounds described in this section correspond to the types $(\mu\text{-H})Os_3(CO)_9(\mu_2\text{-}\eta^1\text{-NHR})^2D$ ($^2D = NCCH_3$ or $N(CH_3)_3$), $(\mu\text{-H})Os_3(CO)_{10}(\mu_2\text{-}\eta^1\text{-NHR})$ (Formula I), and $Os_3(CO)_9(\mu_2\text{-}\eta^1\text{-NHC}_6H_5)(\mu\text{-X})$ (Formula II; X=Cl or I) with endo or exo orientation of the C_6H_5 moiety with respect to the $Os(CO)_4$ group. Only $[(\mu\text{-H})_2Os_3(CO)_{10}(\mu_2\text{-}\eta^1\text{-NHC}_6H_5)][CF_3CO_2]$ (No. 16) has two hydride ligands, bridging two different Os-Os bond edges.

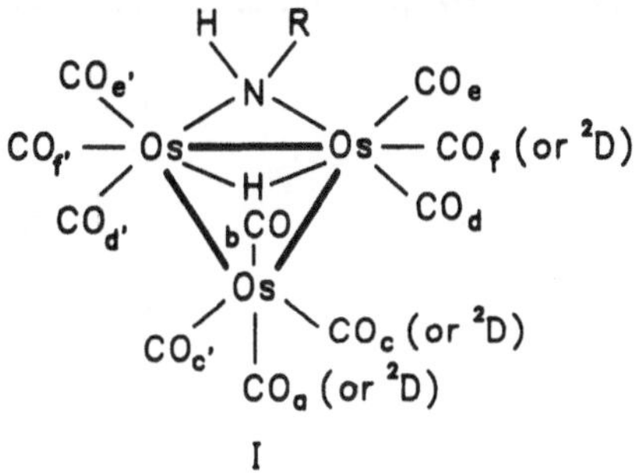

I

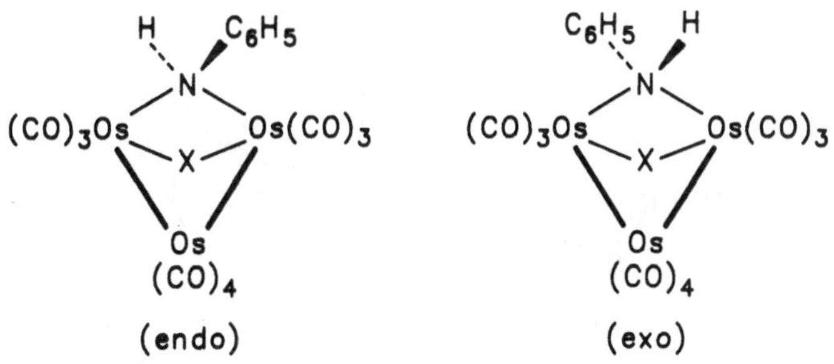

II

The tetracoordinated bridgehead N atoms are formally sp³ hybridized and act as three-electron donors [18, 20].

For $(\mu$-H)Os$_3$(CO)$_9(\mu_2$-η^1-NHCH$_3$)NCCH$_3$ (No. 2; Formula I), a mixture of three isomers was observed. Structures were postulated for Isomers 1 and 3 bearing the NCCH$_3$ group on a bridgehead Os center trans to the μ-H ligand or in an axial position on the unique Os center on the same side of the Os$_3$ triangle as the bridging hydride, respectively, based on ^{1}H NMR spectroscopy; no structure proposal was given for Isomer 2 [30].

X-ray determinations of the compounds of type I show that the dibridged Os–Os distances are approximately 0.02 to 0.11 Å shorter than the nonbridged Os–Os distances ranging between 2.81 to 2.91 Å and which differ only slightly. Only for one of the two crystallographically independent molecules of $(\mu$-H)Os$_3$(CO)$_{10}(\mu_2$-η^1-NHC$_6$F$_5$) (No. 19) is the bridged Os(2)–Os(3) distance of 2.81(3) Å slightly larger than the nonbridged Os(1)–Os(2) bond edge of 2.77(2) Å [32]. A net shortening of the dibridged Os(μ-H)(μ-N)Os linkage relative to nonbridged Os–Os bonds is characteristic for Os(μ-H)(μ-X)Os species containing X atoms with small covalent radii such as H, N, or Cl [8, 10]. On the contrary, for compounds of type II, having an Os(μ-N)(μ-Cl)Os linkage, the dibridged Os–Os distance increases to 3.15 Å (see Nos. 31 and 32). Therefore, the metal–metal framework of compounds of Formula II is reasonably described as an open Os$_3$ triangle, analogously to Os$_3$(CO)$_{10}(\mu$-Cl)$_2$ (see "Organoosmium Compounds" B 3, 1994, p. 71) with a bridging Os–Os separation of 3.159 Å [34].

For the compounds of the type $(\mu$-H)Os$_3$(CO)$_{10}(\mu_2$-η^1-NHN=CR^{1}R^2) (Nos. 25 to 28) the existence of four isomeric configurations, Formulas IIIa to IIId (CO groups omitted), is in principle possible, assuming that the N–H vector points away from the Os$_3$ plane. The isomeric forms IIIa and IIIc, and IIIb and IIId, are interconvertible by rotation about the N–CR^{1}R^2 bond, whereas IIIa and IIIb, and IIIc and IIId, are related by rotation about the N–N bond. A mixture of isomers corresponding to the isomeric forms IIIa and IIIc is only

III

observed for No. 27 ($R^1 = CH_3$, $R^2 = C_6H_5$) in the ^1H NMR spectra, exhibiting two CH_3 and two sets of hydride signals, but no different isomers could be identified for the other compounds of this type. Interestingly, for Nos. 27 and 28, the isomeric forms IIIa and/or IIIc were found in the solid state for both compounds (see Fig. 6 and Fig. 7, pp. 21/2); it appears that in isomers of type IIIb and IIId there would be a very short contact of ca. 1 Å between the hydride ligand and an R group on the diazoalkane ligand. It is probable that there is a high barrier to rotation about the N–N bond and that the isomers IIIb and IIId do not exist even in solution [14].

Many compounds listed in Table 1 were prepared by the following methods:

Method I: Compounds of the type $(\mu\text{-}H)Os_3(CO)_{10}(\mu_2\text{-}\eta^1\text{-}NHR)$ (Formula I) were prepared by reaction of $Os_3(CO)_{12}$ with primary amines NH_2R:

   a. In decane at 145 °C for 7 h or at 170 °C for ca. 4 h ($R = C_6H_5$ and C_6F_5, respectively) [32] or without a solvent at 180 °C for 75 min ($R = C_6H_4F\text{-}4$) or at 186 to 189 °C for 7 h ($R = C_6H_4CH_3\text{-}4$) [1]. The products were purified by TLC with $CHCl_3$/pentane [1] or by column chromatography with $CHCl_3$/hexane [32] as eluants.

   b. In refluxing octane for 7 h ($R = CH_2C_6H_5$) [5], or without a solvent at 184 °C ($R = C_6H_5$) [1] in the presence of CO [1, 5]; workup as before [1]. A reaction time of only 3.5 h reduced the product yield by more than 50% [5].

Method II: Compounds of the type $(\mu\text{-}H)Os_3(CO)_{10}(\mu_2\text{-}\eta^1\text{-}NHR)$ (Formula I) were prepared from $(\mu\text{-}H)_2Os_3(CO)_{10}$ [8, 15] or $[(\mu\text{-}H)Os_3(CO)_{11}]^-$ [18] by treatment with azides RN_3. Preparations with $PO(OC_6H_5)_2N_3$ or 2-chloropyridine-2-azide were performed in refluxing hexane for 20 h, whereas those with an excess of $4\text{-}CH_3C_6H_4SO_2N_3$ [8, 15] or $(CH_3)_3SiN_3$ [18] were conducted in $CHCl_3$ or CH_2Cl_2 at room temperature for 24 h; workup by TLC with hexane or CH_2Cl_2/hexane (1:1) as eluant [15, 18], or by extraction of the residue with ether and crystallization at −25 °C [8]. The reactions probably proceeded via compounds of the type **$(\mu\text{-}H)Os_3(CO)_{10}(\mu_2\text{-}\eta^2\text{-}NHN\text{=}NR)$** (Section 3.1.1.7.3.2) [8, 15].

Method III: Compounds of the type $(\mu\text{-}H)Os_3(CO)_{10}(\mu_2\text{-}\eta^1\text{-}NHR)$ (Formula I; $R = H$, $n\text{-}C_4H_9$, C_6H_5) were prepared from the cyclohexa-1,3-diene complex $Os_3(\eta^4\text{-}C_6H_8)(CO)_{10}$ and an excess of NH_2R in refluxing cyclohexane for 2 to 3 h, followed by TLC purification with CH_2Cl_2/hexane (1:9) as eluant [2].

Method IV: Compounds of the type $(\mu\text{-}H)Os_3(CO)_{10}(\mu_2\text{-}\eta^1\text{-}NHN\text{=}CR^1R^2)$ (Formula I) were prepared from $(\mu\text{-}H)_2Os_3(CO)_{10}$ and 3.5 to 100 equivalents of $CR^1R^2N_2$. The reactions were performed in hexane or ether at 0 °C, ca. 20 °C, or at reflux temperature for 1 to 48 h, depending on the nature of R^1 and R^2, and workup by column chromatography or by TLC with hexane as eluant. Only No. 26 was precipitated directly at −78 °C [14].

Compounds of the type **$(\mu\text{-}H)Os_3(CO)_{10}(\mu_2\text{-}\eta^1\text{-}NHCOX)$** were obtained by treatment of $(\mu\text{-}H)Os_3(CO)_{10}(\mu_2\text{-}\eta^1\text{-}NCO)$ with HX (X = OR or NR^1R^2) [16]; for X = NR^1R^2, see No. 12 and No. 13.

References on pp. 24/5

Table 1

Compounds with μ_2-η^1-Bonded NHR Ligands.

An asterisk preceding the compound number indicates further information at the end of the table, pp. 17/24.

Explanations, abbreviations, and units on p. X.

No. compound	method of preparation (yield) properties and remarks

compounds of the types $(\mu$-H)Os$_3$(CO)$_9(\mu_2$-η^1-NHR)^{2}D and $(\mu$-H)Os$_3$(CO)$_{10}(\mu_2$-η^1-NHR) (Formula I) and $[(\mu$-H)$_2$Os$_3$(CO)$_{10}(\mu_2$-η^1-NHC$_6$H$_5$)][CF$_3$CO$_2$] (No. 16)

1 $(\mu$-H)Os$_3$(CO)$_9(\mu_2$-η^1-NH$_2$)NCCH$_3$ — from No. 4 and (CH$_3$)$_3$NO/CH$_3$CN in THF at 20 °C (50 to 90%) [24]

2 $(\mu$-H)Os$_3$(CO)$_9(\mu_2$-η^1-NHCH$_3$)NCCH$_3$

Isomer 1

Isomer 3

mixture of three isomers, Isomers 1, 2, and 3, differing in the positions of the NCCH$_3$ ligand; no structure was proposed for Isomer 2 (see also p. 2) [30]

from No. 5 and (CH$_3$)$_3$NO/CH$_3$CN in CH$_2$Cl$_2$ at room temperature for 30 min and TLC purification with CH$_2$Cl$_2$/hexane (1:4) giving Isomer 1 (37%), Isomer 2 (27%), and Isomer 3 (25%), together with No. 3 (only 4%) [30]

^{1}H NMR (CDCl$_3$):

Isomer 1: −13.59 (d, OsH; J(H,H)=2.4), 2.48 (s, NCCH$_3$), 2.90 (d, NCH$_3$), 3.67 (br, NH);

Isomer 2: −19.48 (d, OsH), 2.68 (s, NCCH$_3$), 3.41 (d, NCH$_3$; J(H,H)=6.2), 4.22 (br, NH);

Isomer 3: −14.02 (d, OsH; J(H,H)=2.7), 2.54 (s, NCCH$_3$), 3.39 (d, NCH$_3$; J(H,H)=6.1), 4.15 (br, NH) [30]

IR (cyclohexane):

Isomer 1: 1943 m, 1970 ms, 1978 ms, 1995 s, 2003 s, 2013 s, 2049 s, 2091 m (all CO);

Isomer 2: similar to that of Isomer 1, indicating that the NCCH$_3$ ligand is also coordinated to an Os-bridgehead center;

Isomer 3: 1955 m, 1972 s, 1984 s, 2021 s, 2047 s (all CO) [30]

mass spectrum (Isomers 1, 2, and 3): [M]$^+$ [30]

Isomer 1 decomposes slowly in the absence of CH$_3$CN, whereas Isomer 2 is reasonably stable [30]

treatment of Isomers 1, 2, and 3 with P(OCH$_3$)$_3$ in cyclohexane gave the appropriate isomers of $(\mu$-H)Os$_3$(CO)$_9$P(OCH$_3$)$_3(\mu$-NHCH$_3$) ("Organoosmium Compounds" B 4b, in preparation); prolonged heating of Isomer 3 in the presence of P(OCH$_3$)$_3$ led to two isomers of $(\mu$-H)Os$_3$(CO)$_8${P(OCH$_3$)$_3$}$_2$-(μ-NHCH$_3$) (Section 3.1.1.8.3) [30]

References on pp. 24/5

Table 1 (continued)

No. compound	method of preparation (yield) properties and remarks
	treatment with $H_2Os(CO)_4$ (dissolved in heptane; see "Organoosmium Compounds" A 2, 1993, p. 194) in CH_2Cl_2 yielded 39% of No. 5 and 32% of $(\mu-H)_2HOs_4(CO)_{13}$- $(\mu-NHCH_3)$ (see "Organoosmium Compounds" B 8, 1995, Section 4.3.1) [29]
3 $(\mu-H)Os_3(CO)_9(\mu_2-\eta^1-NHCH_3)N(CH_3)_3$	for formation, see No. 2 [30] ^{1}H NMR $(CDCl_3$: -14.28 (d, OsH; J(H,H)$=2.2$), 3.23 (s, N(CH_3)_3), 3.33 (d, NCH_3; J(H,H)$=6.1$), 3.71 (br, NH); the N(CH_3)_3 unit is coordinated to the nonbridged Os center, probably in an equatorial position [30] IR (cyclohexane): 1923 w, 1942 m, 1972, 1997 s, 2011 s, 2021 s, 2047 s, 2081 w (all CO) [30]
4 $(\mu-H)Os_3(CO)_{10}(\mu_2-\eta^1-NH_2)$	III (20%) [2] by thermolysis of $(\mu-H)Os_3(CO)_{10}(\mu_2-\eta^2-$ NHN=NH) (Section 3.1.1.7.3.2) in refluxing hydrocarbons; workup by chromatography (33%), along with unreacted starting material [18, 27]; similarly from $(\mu-H)Os_3(CO)_{10}$- $(\mu_2-\eta^2-NHCH=O)$ (Section 3.1.1.7.3.2) in refluxing nonane for 40 h (50%) [25] from $Os_3(CO)_{11}NH_3$ (see "Organoosmium Compounds" B 3, 1994, p. 254) in trans-decahydronaphthalene (38%); the yield increased to 60% under an NH_3 atmosphere [9] from $Os_3(CO)_{12}$ and an excess of $N(CH_3)_3$ in hexane at ca. 170 °C for 9 h heated in a sealed tube, followed by chromatography with pentane (traces), along with small amounts of $(\mu-H)Os_3\{\mu_2-\eta^1-C=N(CH_3)_2\}$- $(CO)_{10}$ (see "Organoosmium Compounds" B 6, 1993, p. 43), unreacted $Os_3(CO)_{12}$, and traces of $(\mu-H)Os_3(\mu_2-\eta^2-CH=NCH_3)(CO)_{10}$ (see "Organoosmium Compounds" B 5, 1994, p. 189) [4] from $(\mu-H)_2Os_3(CO)_9(\mu_3-\eta^1-NH)$ (Section 3.1.1.7.2.4) by carbonylation in heptane at 140 °C for more than 30 h under ca. 20 atm of CO; purified by chromatography [27] from $(\mu-H)Os_3(CO)_{10}(\mu_2-\eta^1-NO)$ (Section 3.1.1.7.2.2) by hydrogenation in heptane at

References on pp. 24/5

Table 1 (continued)

No. compound	method of preparation (yield) properties and remarks
4 (continued)	140 °C for 2 h with 136 atm of H_2, followed by chromatography with hexane as eluant (26%), along with $(\mu-H)_3HOs_3(CO)_8$-$(\mu_3-\eta^1-NH)$ (30%), $(\mu-H)_2Os_3$-$(CO)_9(\mu_3-\eta^1-NH)$ (10%; both Section 3.1.1.7.2.4), and $(\mu-H)_4Os_4(CO)_{12}$ (26%) (see "Organoosmium Compounds" B 8, 1995, Section 4.1.1.3.1) [27] ^{1}H NMR (no medium given): -15.39 (d, OsH, J(H,H)=3.4), 2.30, 2.92 (br s's, each 1 H) [27] ^{1}H NMR (CDCl$_3$): -15.42 (d, OsH; J(H,H)=3.1; coupling with only one NH$_2$ proton); NH$_2$ signal was not observed in CDCl$_3$ or acetone-d$_6$ [4, 9]; earlier data of -15.25 (s, OsH), 3.75 (br, NH$_2$) in CDCl$_3$ [2] could not be confirmed by [4, 9] ^{1}H NMR (CD$_2$Cl$_2$): -15.4 (OsH) [25] IR (hexane): 1984 w, 1998 s, 2008 s, 2023 s, 2036 w, 2053 m, 2070 s, 2082 w, 2108 w (all CO); (CHCl$_3$): 3349 vw, 3409 vw (NH) [27]; similar data in cyclohexane [2, 4, 25] and pentane [9] mass spectrum: [M]$^+$ [4] structure established by X-ray analysis [25, 26]; no data given decomposed in refluxing octane within 24 h; expected $(\mu-H)_2Os_3(CO)_9(\mu_3-\eta^1-NH)$ (Section 3.1.1.7.2.4) could not be identified [27] hydrogenation in heptane at 140 °C with 136 atm of H_2 gave small amounts of $(\mu-H)_4Os_4(CO)_{12}$ [27] reaction with $(CH_3)_3NO/CH_3CN$ in THF at 20 °C gave No. 1 [24]
5 $(\mu-H)Os_3(CO)_{10}(\mu_2-\eta^1-NHCH_3)$	from $(\mu-H)Os_3(CO)_{10}(\mu-Cl)$ (see "Organoosmium Compounds" B 3, 1994, p. 82) and NH$_2$CH$_3$ gas in refluxing cyclohexane for 10 min; purified by TLC with hexane as eluant (67%) [30] from No. 2 and $H_2Os(CO)_4$ (dissolved in heptane; see "Organoosmium Compounds" A 2, 1993, p. 194) in CH$_2$Cl$_2$ at room temperature for 30 min; purified by TLC with CH$_2$Cl$_2$/hexane (3:7) as eluant (39%), along with 32% $(\mu-H)_2HOs_4(CO)_{13}(\mu-NHCH_3)$ (see "Organoosmium Compounds" B 8, 1995, Section 4.3.1) [29]

Table 1 (continued)

No. compound	method of preparation (yield) properties and remarks
	from $Os_3(CO)_{10}(NCCH_3)_2$ (see "Organoosmium Compounds" B 3, 1994, p. 218) and $NH(CH_3)C_2H_5$ stirred in the dark for 24 h (8%), together with several other products [33]

by decarbonylation of $(\mu-H)Os_3(\mu_2-\eta^2-O=CNHCH_3)(CO)_{10}$ (see "Organoosmium Compounds" B 5, 1994, p. 110) at 125 °C [12]

^{1}H NMR (no medium given): -14.82 (d, OsH; $J(H,H)=2.8$, $J(Os,H)=34$), 3.34 (d, NCH_3; $J(H,H)=6.2$), 3.99 (br, NH) [30]

IR (cyclohexane): 1953 br vw, 1979 m, 1991 s, 2003 s, 2020 s, 2051 s, 2065 s, 2104 mw (all CO) [30]

mass spectrum: $[M]^+$ [30]

decarbonylation at 150 °C yielded $(\mu-H)_2Os_3(CO)_9(\mu_3-\eta^1-NCH_3)$ (Section 3.1.1.7.2.4) [12]

treatment with $(CH_3)_3NO/CH_3CN$ in CH_2Cl_2 yielded No. 2 as a mixture of three isomers, along with No. 3 in low yield [30]

*6 $(\mu-H)Os_3(CO)_{10}(\mu_2-\eta^1-NHCH_2CF_3)$

by hydrogenation of $(\mu-H)Os_3(\mu_2-\eta^2-NH=CCF_3)(CO)_{10}$ (see "Organoosmium Compounds" B 5, 1994, p. 192) in hexane at 120 °C for 16 h under 23 atm of H_2 in an autoclave; workup by TLC with CH_2Cl_2/hexane (1:9) as eluant (80%), along with two minor not further characterized by-products [19]

by hydrogenation of $(\mu-H)Os_3(CO)_{10}(\mu_2-\eta^1-N=CHCF_3)$ (Section 3.1.1.7.2.3) in hexane at 140 °C for 16 h under 49 atm of H_2 as before (30%), along with 25% of $(\mu-H)_2Os_3(CO)_9(\mu_3-\eta^1-NCH_2CF_3)$ and 20% of $(\mu-H)_3HOs_3(CO)_8(\mu_3-\eta^1-NCH_2CF_3)$ (both Section 3.1.1.7.2.4) [13, 19]

yellow crystals (from hexane) [19]

^{1}H NMR (CD_2Cl_2): -14.84 (s, OsH), 4.18 (m, CH_2), 7.65 (s, NH) [13, 19]

IR (hexane): 1981 m, 1993 s, 1997 sh, 2005 s, 2008 sh, 2023 s, 2033 m, 2054 s, 2069 vs, 2016 w (all CO) [13, 19]; (Nujol mull): 1156 vs, 1165 vs, 1206 s, 1269 m, 1280 s, 1288 m (all CF), 3375 w (NH) [19]

EI mass spectrum: $[M]^+$ [19]

References on pp. 24/5

Table 1 (continued)

No. compound	method of preparation (yield) properties and remarks

7 $(\mu\text{-H})Os_3(CO)_{10}(\mu_2\text{-}\eta^1\text{-NHC}_3H_7\text{-n})$

from $Os_3(CO)_{10}(NCCH_3)_2$ and an excess of $NH(C_2H_5)C_3H_7\text{-n}$ in CH_2Cl_2 at room temperature for 24 h; purified by TLC with CH_2Cl_2/hexane (3:10) as eluant (13%), along with 15% of $(\mu\text{-H})Os_3\{\mu_2\text{-}\eta^1\text{-}CHCH=N(C_2H_5)C_3H_7\text{-n}\}(CO)_{10}$ (see "Organoosmium Compounds" B 6, 1993, p. 11) [33]

yellow crystals (from hexane at −20 °C) [33]

^{1}H NMR (CDCl$_3$): −14.92 (d, OsH; ^{3}J(H,H)=3.3), 0.95 (t, CH$_3$; J(H,H)=3.3), 1.58, 2.80 (m's, each 2 H), 4.13 (br s, NH) [33]

IR (hexane): 1976 w, 1988 s, 1998 sh, 2000 s, 2019 vs, 2048 s, 2062 vs, 2100 w (all CO) [33]

8 $(\mu\text{-H})Os_3(CO)_{10}(\mu_2\text{-}\eta^1\text{-NHC}_4H_9\text{-n})$

III (40%) [2]

^{1}H NMR (CDCl$_3$, 35 °C): −14.62 (s, OsH), 2.75 (m, CH$_2$CH$_2$CH$_3$), 4.0 (br, NCH$_2$); no resonance for NH given [2]

^{13}C NMR (toluene-d$_8$; −90 to +70 °C): 172.8, 174.9 (J(C,C)=12.2), 178.2, 179.4 (each 2 CO, Os(CO)$_3$, and equatorial CO of Os(CO)$_4$), 183.4, 191.3 (each 1 CO, axial CO of Os(CO)$_4$); no resonances for n-C$_4$H$_9$ given; the static structure cannot persist in solution; No. 8 is diastereotopic but no evidence for the existence of isomers could be established even at −90 °C [6]

IR (cyclohexane): 1978 w, 1989 m, 2002 s, 2021 s, 2050 m, 2065 s, 2103 w (all CO) [2]

9 $(\mu\text{-H})Os_3(CO)_{10}(\mu_2\text{-}\eta^1\text{-NHCH}_2CO_2C_2H_5)$

from $(\mu\text{-H})Os_3(CO)_{10}(\mu\text{-OH})$ (see "Organoosmium Compounds" B 3, 1994, p. 94) and $NH_2CH_2CO_2C_2H_5$ at 70 °C (36%) [20]

10 $(\mu\text{-H})Os_3(CO)_{10}\{\mu_2\text{-}\eta^1\text{-NHCH}(CH_3)CO_2C_2H_5\}$

from $(\mu\text{-H})Os_3(CO)_{10}(\mu\text{-OH})$ and $NH_2CH(CH_3)CO_2C_2H_5$ at 70 °C (61%) [20]

11 $[(\mu\text{-H})Os_3(CO)_{10}\{\mu_2\text{-}\eta^1\text{-NHCH}=N(CH_3)_2\}][CF_3CO_2]$

by protonation of $(\mu\text{-H})Os_3(CO)_{10}\{\mu_2\text{-}\eta^1\text{-}N=CHN(CH_3)_2\}$ (Section 3.1.1.7.2.3) with CF_3CO_2H in CD_2Cl_2 at −60 °C, based on ^{1}H NMR spectra (not isolated) [22]

^{1}H NMR (CD$_2$Cl$_2$, −60 °C): −12.35 (OsH), 2.68, 3.16 (s's, each 1 CH$_3$), 5.20 (d, NH; J(CH,NH)=18.1), 7.07 (d, CH); nonequiva-

Table 1 (continued)

No. compound	method of preparation (yield) properties and remarks

lence of the CH_3 groups resulted from the C–N double bond [22]

decomposed upon warming to room temperature with slow formation of $(\mu\text{-}H)Os_3(CO)_{10}$-$(\mu_2\text{-}\eta^2\text{-}O_2CCF_3)$ (see "Organoosmium Compounds" B 3, 1994, p. 125), based on ^1H NMR spectra [22]

12 $(\mu\text{-}H)Os_3(CO)_{10}(\mu_2\text{-}\eta^1\text{-}NHCONHNH_2)$ from $(\mu\text{-}H)Os_3(CO)_{10}(\mu_2\text{-}\eta^1\text{-}NCO)$ (Section 3.1.1.7.2.3) and N_2H_4 in CH_2Cl_2 at room temperature (quantitative) [16]

from $Os_3(CO)_{12}$, $Os_3(CO)_{11}NCCH_3$, or $Os_3(CO)_{10}(NCCH_3)_2$ with N_2H_4 neat or in CH_2Cl_2 (5 to 10%) [16]

^1H NMR ($CDCl_3$): -14.64 (d, OsH), 3.93 (s, NH_2), 6.84 (s, NH), 7.25 (br s, NH) [16]

IR (KBr): 1695 (CO of $NHCONHNH_2$); (cyclohexane): 1983 w, 2001 s, 2007 s, 2020 s, 2056 s, 2071 w, 2107 w (all CO) [16]

mass spectrum: $[M]^+$ [16]

structure established by X-ray analysis (no details given) [16, 17]

13 $(\mu\text{-}H)Os_3(CO)_{10}\{\mu_2\text{-}\eta^1\text{-}NHCONHN(CH_3)_2\}$

from $(\mu\text{-}H)Os_3(CO)_{10}(\mu_2\text{-}\eta^1\text{-}NCO)$ (Section 3.1.1.7.2.3) and $NH_2N(CH_3)_2$ in CH_2Cl_2 (quantitative) [16]

from $Os_3(CO)_{12}$, $Os_3(CO)_{11}NCCH_3$, or $Os_3(CO)_{10}(NCCH_3)_2$ with $NH_2N(CH_3)_2$ neat or in CH_2Cl_2 (5 to 10%) [16]

^1H NMR ($CDCl_3$): -14.67 (d, OsH), 2.58 (s, CH_3), 6.32 (s, NH), 7.32 (br s, NH) [16]

IR (KBr): 1700 (CO of $NHCONHN(CH_3)_2$); (cyclohexane): 1984 w, 2001 s, 2008 sh, 2021 s, 2058 s, 2072 s, 2108 w (all CO) [16]

mass spectrum: $[M]^+$ [16]

structure established by X-ray analysis [16, 17]

14 $(\mu\text{-}H)Os_3(CO)_{10}(\mu_2\text{-}\eta^1\text{-}NHCH_2C_6H_5)$ Ib (16%), along with $(\mu\text{-}H)Os_3(\mu_2\text{-}\eta^2\text{-}O{=}CNHCH_2C_6H_5)(CO)_{10}$ (see "Organoosmium Compounds" B 5, 1994, p. 112) as the main product [5]

from $Os_3(CO)_{12}$ and an excess of C_6H_5CN in refluxing octane for 6 h under H_2 in the presence of an excess of CH_3CO_2H; purified by TLC with petroleum ether/ether (19:1) as eluant (high yield), together with

References on pp. 24/5

Table 1 (continued)

No. compound	method of preparation (yield) properties and remarks
14 (continued)	$(\mu-H)Os_3(CO)_{10}(\mu_2-\eta^1-N=CHC_6H_5)$ (Section 3.1.1.7.2.3) and $(\mu-H)Os_3(CO)_{10}(\mu_2-\eta^2-O_2CCH_3)$ (see "Organoosmium Compounds" B 3, 1994, p. 124); the yield decreased considerably in the absence of CH_3CO_2H and/or H_2 or with shorter reaction times; similar results were obtained starting from $(\mu-H)_2Os_3(CO)_{10}$ instead of $Os_3(CO)_{12}$ [23]
	from $Os_3(cyclo-C_8H_{14})_2(CO)_{10}$ and $NH_2CH_2C_6H_5$ in cyclohexane/C_8H_{14} at 60 to 70 °C for 2 h, followed by TLC with petroleum ether/$CHCl_3$ as eluant (3%), along with small amounts of $(\mu-H)Os_3(\mu_2-\eta^2-O=CNHCH_2C_6H_5)(CO)_{10}$ (see "Organoosmium Compounds" B 5, 1994, p. 112) [5]
	by thermolysis of $(\mu-H)Os_3(\mu_2-\eta^2-O=CNHCH_2C_6H_5)(CO)_{10}$ in d_8-toluene at 150 °C for 47 h in a sealed NMR tube (14%), along with $Os_3(CO)_{12}$, based on ^1H NMR spectra; similarly by thermolysis in the presence of a fivefold excess of benzylamine (31%), or by thermolysis in decane for 1 h under 1 atm of H_2 (18%); using CO instead of H_2 led only to trace amounts [5]
	yellow crystals (from petroleum ether/$CHCl_3$), m.p. 157 to 159 °C [5]
	^1H NMR ($CDCl_3$): -14.29 (d, OsH; J(H,H) = 2.7), 4.01 (d, CH_2), 4.58 (br, NH), 7.2 to 7.4 (m, C_6H_5) [23]; similar data in [5]
	IR (hexane): 1978 sh w, 1981 w, 1991 s, 2003 s, 2021 vs, 2051 s, 2067 vs, 2104 w (all CO) [23]; similar data in [5]
*15 $(\mu-H)Os_3(CO)_{10}(\mu_2-\eta^1-NHC_6H_5)$	Ia [32], Ib (70%) [1]
	III (50%), along with 40% of $(\mu-H)_2Os_3(\mu_3-\eta^2-NHC_6H_4)(CO)_9$ (see "Organoosmium Compounds" B 5, 1994, p. 285) [2]
	from $(\mu-H)Os_3(CO)_{10}(\mu-OH)$ (see "Organoosmium Compounds" B 3, 1994, p. 94) and $NH_2C_6H_5$ at 75 °C (no further details given) [21]
	by carbonylation of $(\mu-H)Os_3(\mu_2-\eta^2-NHC_6H_4)-(CO)_9$ or $(\mu-H)_2Os_3(\mu_2-\eta^2-NHC_6H_4)-(CO)_8NH_2C_6H_5$ (both "Organoosmium Compounds" B 5, 1994, p. 285) in refluxing octane (43%) [1]

References on pp. 24/5

Table 1 (continued)

No. compound	method of preparation (yield) properties and remarks

by carbonylation of $(\mu-H)_2Os_3(CO)_9(\mu_3-\eta^1-NC_6H_5)$ (Section 3.1.1.7.2.4) at 195 °C for 10 h under 1 atm of CO (50%); yields of only 10 to 20% at 175 °C, no formation at 150 °C [1]

by-product in the $Os_3(CO)_{12}$-catalyzed conversion of $NH(CH_3)C_6H_5$ to $CH_2\{C_6H_4NHCH_3-4\}_2$

by reaction of $NH_2C_6H_5$ generated in situ with $Os_3(CO)_{12}$ under the conditions of Preparation Method Ib (8%); similar reaction in octane at 140 to 170 °C for 100 h under vacuum in a sealed tube gave the title compound exclusively but in a yield of only 4% [7]

orange crystals (from pentane), m.p. 185 to 187 °C [7]

^1H NMR $(CDCl_3)$: -14.13 (d, OsH; J(H,H)=3.1), 5.8 (br, NH), 6.7 to 7.3 (m, C_6H_5) [1]

^1H NMR $(CF_3CO_2H/CH_2Cl_2$ $\{1:8\})$: -14.1 (d, OsH; ^3J=3.1), indicating that the protonation proceeded at slow rates [31]

UV $(CH_2Cl_2;$ ε): 283 (sh; 9100), 325 (6800), 405 (2300) [31]

IR (cyclohexane): 1976 w, 1984 m, 1995 s, 2006 s, 2023 s, 2053 s, 2069 s, 2106 m (all CO), 3360 (NH) [1]

*16 $[(\mu-H)_2Os_3(CO)_{10}(\mu_2-\eta^1-NHC_6H_5)][CF_3CO_2]$

by protonation of No. 15 with CF_3CO_2H in CH_2Cl_2 [31]

^1H NMR $(CF_3CO_2H/CH_2Cl_2$ $\{1:8\})$: -18.2 (OsH), -14.7 (d, OsH; ^3J=1.1) [31]

UV (CF_3CO_2H/CH_2Cl_2): 308 sh, 366 [31]

17 $(\mu-H)Os_3(CO)_{10}\{\mu_2-\eta^1-NHC_6H_5Cr(CO)_3\}$

from $(\mu-H)Os_3(CO)_{10}(\mu-OH)$ (see "Organoosmium Compounds" B 3, 1994, p. 94) and $NH_2C_6H_5Cr(CO)_3$ in decane at 150 °C (no further data given) [21]

18 $(\mu-H)Os_3(CO)_{10}(\mu_2-\eta^1-NHC_6H_4F-4)$

Ia (41%), along with 12% of $(\mu-H)_2Os_3(\mu_3-\eta^2-NHC_6H_3F-5)(CO)_9$ and 13% of $(\mu-H)_2Os_3-(\mu_3-\eta^2-NHC_6H_3F-5)(CO)_8NH_2C_6H_4F-4$ (both "Organoosmium Compounds" B 5, 1994, p. 286) [1]

orange solid (from CH_2Cl_2/pentane), m.p. 194 to 196 °C [1]

^1H NMR $(CDCl_3)$: -12.51 (d, OsH; J(H,H)=2.8), 5.7 (br, NH), 6.6 to 7.3 (m, C_6H_4) [1]

References on pp. 24/5

Table 1 (continued)

No. compound	method of preparation (yield) properties and remarks
18 (continued)	IR (cyclohexane): 1976 m, 1985 m, 1995 s, 2006 s, 2024 s, 2054 s, 2069 s, 2106 m (all CO) [1]

18 (continued) — thermolysis in refluxing nonane gave $(\mu-H)_2Os_3(\mu_3-\eta^2-NHC_6H_3F-5)(CO)_9$ (see "Organoosmium Compounds" B 5, 1994, p. 286) exclusively but in octane at 186 to 194 °C (sealed tube) a mixture with $(\mu-H)_2Os_3(CO)_9(\mu_3-\eta^1-NC_6H_4F-4)$ (Section 3.1.1.7.2.4) was obtained [1]

*19 $(\mu-H)Os_3(CO)_{10}(\mu_2-\eta^1-NHC_6F_5)$

Ia (38%) [32]
orange crystals (from hexane/CHCl_3) [32]
^1H NMR (CDCl_3): -13.77 (m, OsH), 5.8 (s, NH) [32]
IR (cyclohexane): 1986 w, 1999 s, 2008 m, 2024 s, 2055 m, 2070 s, 2108 w (all CO) [32]
thermolysis in refluxing decane yielded $(\mu-H)_2Os_3(\mu_3-\eta^2-NHC_6F_4)(CO)_9$ (see "Organoosmium Compounds" B 5, 1994, p. 287) in 45% yield [32]

20 $(\mu-H)Os_3(CO)_{10}(\mu_2-\eta^1-NHC_6H_4CH_3-4)$

Ia (65%) [1]
orange crystals (from pentane), m.p. 166 °C [1]
^1H NMR (CDCl_3): -14.15 (d, OsH; J(H,NH) = 3.1), 2.26 (s, CH_3), 5.8 (br, NH), 6.69, 7.01 (m's, C_6H_4) [1]
IR (cyclohexane): 1976 m, 1984 m, 1995 s, 2005 s, 2023 s, 2035 w, 2053 s, 2068 s, 2104 m (all CO) [1]
thermolysis at ca. 210 °C for 4 h in a sealed tube gave only small amounts of $(\mu-H)_2Os_3(CO)_9(\mu_3-\eta^1-NC_6H_4CH_3-4)$ (Section 3.1.1.7.2.4); no decomposition occurred in refluxing nonane [1]

21 $(\mu-H)Os_3(CO)_{10}(\mu_2-\eta^1-NHC_6H_3Cl_2-2,4)$

no preparation reported
thermolysis in refluxing nonane for 3 h yielded $(\mu-H)_2Os_3(\mu_3-\eta^2-NHC_6H_2Cl_2-3,5)(CO)_9$ (see "Organoosmium Compounds" B 5, 1994, p. 287) [32]

22 $(\mu-H)Os_3(CO)_{10}(\mu_2-\eta^1-NHNC_5H_3Cl-6)$
   $NC_5H_3Cl-6 = 6$-chloropyridin-2-yl

II (96%) [15]
^1H NMR (acetone-d_6): -14.18 (d, OsH; J=2), 6.12 (br, NH), 7.10, 7.17 (d's, each 1 H; J=12), 7.68 (dd, 1 H) [15]

References on pp. 24/5

Table 1 (continued)

No. compound	method of preparation (yield) properties and remarks

IR (hexane): 1970 w, 1985 w, 2000 s, 2006 m, 2022 s, 2056 s, 2072 s, 2105 w (all CO) [15]

mass spectrum: $[M]^+$ [15]

decarbonylation in hexane at 130 °C for 14 h under 5 atm of Ar yielded $(\mu\text{-H})Os_3(CO)_9$-$(\mu_3\text{-}\eta^2\text{-NHC}_5H_3NCl\text{-}6)$ (Section 3.1.1.7.3.3) [15]

*23 $(\mu\text{-H})Os_3(CO)_{10}(\mu_2\text{-}\eta^1\text{-NHSO}_2C_6H_4CH_3\text{-}4)$

II (39%) [8]

bright yellow crystals (from CH_2Cl_2) [8]

^{1}H NMR (CDCl$_3$): -15.31 (d, OsH), 2.45 (s, CH$_3$), 5.80 (br, NH), 7.39 (d, H-3,5), 7.78 (d, H-2,6); J(H-2,3) = J(H-5,6) = 8.1, J(μ-H,NH) = 2.2 [8]

IR (cyclohexane): 1974 w, 1990 w, 2004 s, 2011 m, 2024 vs, 2060 m, 2078 vs, 2110 vw (all CO); (CCl$_4$): 3325 vw (NH) [8]

*24 $(\mu\text{-H})Os_3(CO)_{10}\{\mu_2\text{-}\eta^1\text{-NHS=N(Si(CH}_3)_3)_2\}$

from $(\mu\text{-H})_2Os_3(CO)_{10}$ and $S\{NSi(CH_3)_3\}_2$ (1:2) in refluxing hexane for 15 h, followed by TLC with cyclohexane/CH$_2$Cl$_2$ as eluant (25%) [28]

orange-yellow, air-stable crystals (from pentane); m.p. 104 °C [28]

^{1}H NMR (CDCl$_3$): -14.80 (d, OsH; ^{3}J(H,NH) = 3.3), 0.28 (s, CH$_3$), 3.93 (br, NH) [28]

^{13}C $\{^1$H$\}$ NMR (CD$_2$Cl$_2$, -80 °C; for numbering, see Formula I, p. 1): 170.1 (d, CO$_{d,d'}$), 174.1 (d, CO$_{e,e'}$), 176.8 (s, CO$_{c,c'}$), 177.3 (s, CO$_{f,f'}$), 181.3 (s, CO$_{b \text{ or } a}$), 189.2 (s, CO$_{a \text{ or } b}$), other resonances omitted; ^{2}J(C,H) = 3.4 or 12.0; spectrum depicted; the ^{13}C NMR data are consistent with the solid state structure [28]

variable-temperature ^{13}C NMR measurements in toluene-d$_8$ indicated pseudo-rotation of the unbridged Os(CO)$_4$ unit above +50 °C as the only fluxional process by coalescence of the signals of CO$_a$, CO$_b$, and CO$_{c,c'}$, resulting in a new signal at ca. 181.5 ppm at +90 °C [28]

IR (cyclohexane): 1973 w, 1982 m, 1995 vs, 2001 vs, 2024 s, 2054 s, 2070 vs, 2105 m (all CO); (KBr?): 757, 817, 846, 877, 898 (NS, SiCH$_3$), 3335 (NH) [28]

mass spectrum: $[M]^+$, $[M-n\,CO]$, n = 1 to 10 [28]

References on pp. 24/5

Table 1 (continued)

No. compound	method of preparation (yield) properties and remarks

25 $(\mu-H)Os_3(CO)_{10}(\mu_2-\eta^1-NHN=CHC_6H_5)$

IV (5%) [14]
^1H NMR $(CDCl_3)$: -14.47 (d, OsH; J(H,NH)= 2.4), 4.29 (br, NH), 5.09 (s, CH), 7.08 to 7.59 (m, C_6H_5) [14]
IR (hexane): 1974 w, 1984 w, 2000 s, 2006 s, 2021 s, 2056 s, 2071 s, 2106 w (all CO) [14]
EI mass spectrum: $[M]^+$ [14]

26 $(\mu-H)Os_3(CO)_{10}\{\mu_2-\eta^1-NHN=C(CH_3)_2\}$

IV (36%) [14]
yellow solid at $-78\,°C$ which darkened at $20\,°C$ in vacuum [14]
^1H NMR (CD_2Cl_2): -14.59 (d, OsH; J(H,NH)= 2.4), 1.88 (s, CH_3), 8.90 (br, NH) [14]
IR (hexane): 1970 s, 1983 s, 1998 s, 2003 m, 2020 s, 2055 s, 2070 s, 2105 w (all CO) [14]
EI mass spectrum: $[M]^+$ [14]
decomposed to $(\mu-H)_2Os_3(CO)_{10}$ in solution at $20\,°C$, possibly by initial conversion into an $NHNC(CH_3)_2$-chelating form, followed by loss of N_2 and a β-hydrogen transfer with elimination of an alkene, but no intermediates were detected [14]

*27 $(\mu-H)Os_3(CO)_{10}\{\mu_2-\eta^1-NHN=C(CH_3)C_6H_5\}$

IV (24%); mixture of isomers; see also p. 2 [14]
yellow hexagonal plates (from hexane) [14]
^1H NMR $(CDCl_3)$: -14.45, -14.59 (d's, OsH; J(H,NH) = 1.6 and 2.1), 1.99, 2.03 (s's, CH_3), 7.72 (br, NH), 6.98 to 7.47 (2 m's, C_6H_5) [14]
IR (hexane): 1972 w, 1985 w, 2000 s, 2004 s, 2022 s, 2057 s, 2071 s, 2106 s (all CO) [14]
EI mass spectrum: $[M]^+$ [14]

*28 $(\mu-H)Os_3(CO)_{10}\{\mu_2-\eta^1-NHN=C(C_6H_5)_2\}$

IV (72%) [14]; see also [3, 10]
by dehydrogenation of $NH_2NHCH(C_6H_5)_2$ with $Os_3(CO)_{10}(NCCH_3)_2$ or $Os_3(cyclo-C_8H_{14})_2(CO)_{10}$ [11]; no further data given in [10]
yellow blocks (from hexane) [14]
^1H NMR $(CDCl_3)$: -14.41 (d, OsH; J(H,NH)= 1.6), 4.24 (br, NH), 7.01 to 7.46 (m, C_6H_5) [14]
IR (hexane): 1979 w, 1997 s, 2001 s, 2017 s, 2054 s, 2067 s, 2104 w (all CO) [14]
EI mass spectrum: $[M]^+$ [14]

References on pp. 24/5

Table 1 (continued)

No. compound	method of preparation (yield) properties and remarks

no decomposition was observed upon thermolysis in heptane at 170 °C for 1 h under
argon [14]

29 $(\mu\text{-H})Os_3(CO)_{10}\{\mu_2\text{-}\eta^1\text{-NHP=O}(OC_6H_5)_2\}$

II (46%) [15]
^{1}H NMR: -14.82 (d, OsH; J=1), 7.24 to 7.70
(m, C_6H_5); no resonance for NH given [15]
IR (hexane): 1970 w, 1989 w, 2004 s, 2014 m,
2025 s, 2061 s, 2078 s, 2111 s (all CO) [15]

*30 $(\mu\text{-H})Os_3(CO)_{10}\{\mu_2\text{-}\eta^1\text{-NHSi}(CH_3)_3\}$

II (main product) [18]
yellow crystals (from hexane?) [18]
^{1}H NMR (no medium given): -12.49 (OsH),
0.9? (br s, NH), 1.28 (CH_3) [18]
IR (no medium given): 1255 ($Si(CH_3)_3$),
1952 vw, 1980 w, 1985 m, 1997 s, 2002 sh w,
2016 vs, 2020 vs, 2056 vs, 2069 vs, 2109 m
(all CO) [18]
mass spectrum: $[M]^+$ [18]

compounds of the type $Os_3(CO)_{10}(\mu_2\text{-}\eta^1\text{-NHC}_6H_5)(\mu\text{-X})$ (Formula II)

*31 endo-$Os_3(CO)_{10}(\mu_2\text{-}\eta^1\text{-NHC}_6H_5)(\mu\text{-Cl})$

by protonation of $[N\{P(C_6H_5)_3\}_2]$-
$[Os_3(CO)_9(\mu_3\text{-}\eta^1\text{-NC}_6H_5)(\mu\text{-Cl})]$ (Section
3.1.1.7.2.4) with CF_3CO_2H in THF at -80 °C
under CO, followed by warmup to room temperature and purification by TLC with
CH_2Cl_2/hexane as eluant (small amounts),
along with No. 32, $(\mu\text{-H})Os_3(CO)_{10}(\mu\text{-Cl})$, and
84% of $(\mu\text{-H})Os_3(CO)_9(\mu_3\text{-}\eta^1\text{-NC}_6H_5)(\mu\text{-Cl})$
(Section 3.1.1.7.2.4) [34]
by-product in the preparation of
$Os_3\{\mu_2\text{-}\eta^2\text{-C(OR)NC}_6H_5\}(CO)_{10}(\mu\text{-Cl})$ (R=H,
CH_3) by treatment of $[N\{P(C_6H_5)_3\}_2]$-
$[Os_3(\mu_2\text{-}\eta^2\text{-C(O)NC}_6H_5)(CO)_{10}(\mu\text{-Cl})]$ (see
"Organoosmium Compounds" B 5, 1994,
pp. 248/51) with CF_3CO_2H or $CF_3SO_3CH_3$ as
described before; protonation yielded 1% of
No. 31 and 4% of No. 32, whereas the
alkylation gave only No. 32 in 4% yield [34]
similar protonations with $HBF_4 \cdot O(C_2H_5)_2$ led to
higher yields of Nos. 31 and 32 [34]
^{1}H NMR (CD_2Cl_2): 5.17 (br, NH), 7.0 to 7.4
(C_6H_5) [34]

References on pp. 24/5

Table 1 (continued)

No. compound	method of preparation (yield) properties and remarks

*31 (continued)

^{13}C NMR (THF-d$_8$, -73 °C): 125 to 131 (C$_6$H$_5$), 152.3 (m, C–1 of C$_6$H$_5$), 168.2, 179.8 (s's, each 2 CO), 180.1 and 180.2 (two overlapping peaks corresponding to 3 CO), 181.2 (s, 1 CO), 183.2 (s, 2 CO) [34]
IR (pentane): 1985 m, 1996 s, 2019 vs, 2057 m, 2071 vs, 2106 vw (all CO) [34]
EI mass spectrum: [M]$^+$ [34]

*32 exo-Os$_3$(CO)$_{10}$(μ_2-η^1-NHC$_6$H$_5$)(μ–Cl)

for formation, see No. 31 [34]
^{1}H NMR (CD$_2$Cl$_2$): 3.64 (br, NH), 7.02 to 7.31 (C$_6$H$_5$) [34]
^{13}C NMR (THF-d$_8$, -73 °C): 118 to 131 (C$_6$H$_5$), 166.8 (t, C–1 of C$_6$H$_5$; ^{3}J(C,H)$=6.4$), 167.7, 177.0, 177.1 (s's, each 2 CO), 179.5 (s, 1 CO), 180.0 (s, 2 CO), 185.1 (s, 1 CO) [34]
IR (pentane): 1973 vw, 1985 w, 2005 vs, 2021 s, 2036 vw, 2058 m, 2074 s, 2106 w (all CO) [34]
EI mass spectrum: [M]$^+$ [34]

33 endo-Os$_3$(CO)$_{10}$(μ_2-η^1-NHC$_6$H$_5$)(μ–I)

by protonation of [N{P(C$_6$H$_5$)$_3$}$_2$][Os$_3$(μ_3–CO)-(CO)$_8$(μ_3-η^1-NC$_6$H$_5$)I] in the presence of minor amounts of [N{P(C$_6$H$_5$)$_3$}$_2$]-[Os$_3$(CO)$_9$(μ_3-η^1-NC$_6$H$_5$)(μ–I)] (both Section 3.1.1.7.2.4) with CF$_3$CO$_2$H in THF at -80 °C overnight in the presence or absence of CO, followed by warmup to room temperature and purification by TLC with CH$_2$Cl$_2$/hexane as eluant (traces to 7%), along with traces to 8% of No. 34, (μ–H)Os$_3$(CO)$_{10}$(μ–I) (see "Organoosmium Compounds" B 3, 1994, p. 84), and 22 to 32% of (μ–H)Os$_3$(CO)$_9$-(μ_3-η^1-NC$_6$H$_5$)(μ–I) (Section 3.1.1.7.2.4) [34]
by–product in the preparation of Os$_3${μ_2-η^2-C(OR)NC$_6$H$_5$}(CO)$_{10}$(μ–I) (R$=$H, CH$_3$) by treatment of [N{P(C$_6$H$_5$)$_3$}$_2$]-[Os$_3$(μ_2-η^2-C(O)NC$_6$H$_5$)(CO)$_{10}$(μ–I)] (see "Organoosmium Compounds" B 5, 1994, pp. 249/52) with CF$_3$CO$_2$H or CF$_3$SO$_3$CH$_3$ as described before; protonation yielded 3% of No. 33 and 5% of No. 34, whereas the alkylation gave only No. 34 in 2% yield [34]
^{1}H NMR (CD$_2$Cl$_2$): 4.73 (br, NH), 7.1 to 7.3 (m, C$_6$H$_5$) [34]
IR (pentane): 1981 w, 1995 m, 2005 m, 2018 vs, 2053 m, 2067 s, 2105 w (all CO) [34]
EI mass spectrum: [M]$^+$ [34]

References on pp. 24/5

Table 1 (continued)

No. compound	method of preparation (yield) properties and remarks
34 exo-Os$_3$(CO)$_{10}$(μ_2-η^1-NHC$_6$H$_5$)(μ-I)	for formation, see No. 33 [34] ^{1}H NMR (CD$_2$Cl$_2$): 3.88 (br, NH), 6.97 to 7.31 (C$_6$H$_5$) [34] IR (pentane): 1973 vw, 1984 w, 1993 w, 2007 vs, 2019 m, 2057 m, 2072 s, 2105 w (all CO) [34] EI mass spectrum: [M−H]$^+$ [34]

*Further information:

(μ-H)Os$_3$(CO)$_{10}$(μ_2-η^1-NHCH$_2$CF$_3$) (Table 1, No. 6) crystallizes in the monoclinic space group C2/c − C$^6_{2h}$ (No. 15) with a = 19.983(3), b = 15.562(3), c = 14.089(3) Å, β = 118.80(3)°; Z = 8, D$_c$ = 3.278 g/cm³. The structure is shown in **Fig. 1**. The three-electron donating μ_2-η^1-NHCH$_2$CF$_3$ ligand is attached to the cluster via the nitrogen atom spanning the Os(1)–Os(3) bond edge. Approximately tetrahedral geometry of both the nitrogen and carbon atom indicates that the organic ligand is saturated; the N–C distance of 1.460 Å is consistent with a single bond. The NH, CH$_2$, and μ-H hydrogen atoms were not located directly [19].

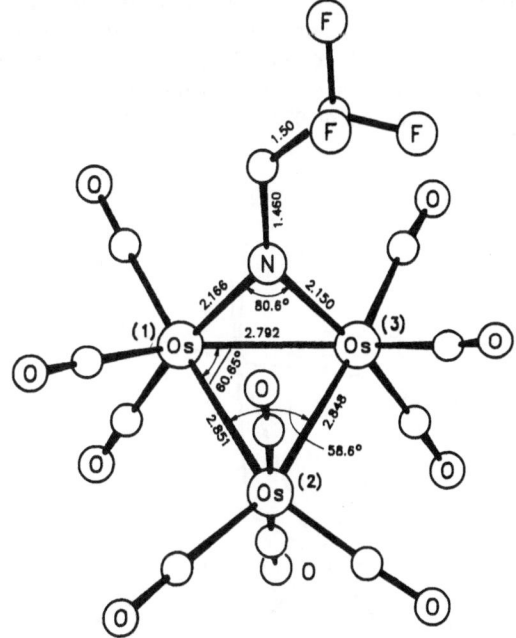

Fig. 1. Molecular structure of (μ-H)Os$_3$(CO)$_{10}$(μ_2-η^1-NHCH$_2$CF$_3$) (No. 6) with selected bond distances (in Å) and bond angles [19].

(μ-H)Os$_3$(CO)$_{10}$(μ_2-η^1-NHC$_6$H$_5$) and [(μ-H)$_2$Os$_3$(CO)$_{10}$(μ_2-η^1-NHC$_6$H$_5$)][CF$_3$CO$_2$] (Table 1, Nos. 15 and 16). Protonation of No. 15 with CF$_3$CO$_2$H led to No. 16 by addition of a proton spanning one of the two equivalent Os(CO)$_3$–Os(CO)$_4$ bond edges in No. 16, as monitored by ^{1}H NMR spectroscopy; the protonation proceeded at slow rates. The ratio of the equilibrium concentration of No. 16 to No. 15 was determined by varying the CF_3CO_2H concentration. The degree of protonation was estimated as 3.0, 1.3, and 0.19 for CF_3CO_2H/CH_2Cl_2 mixtures of 1:6, 1:8, and 1:20 (by volume), respectively, at 23 °C and an initial concentration of No. 15 of 0.062 mol/L. The equilibrium constant [No. 16] /[No. 15] amounts to 440 in CF_3CO_2H/CH_2Cl_2 (1:8) at 23 °C and was calculated from the degree of protonation [31].

The observed hypsochromic shifts of the UV absorptions indicate significant changes in the electronic structure upon protonation of No. 15, increasing the energy difference between the highest occupied and the lowest unoccupied molecular orbitals [31].

Complete deprotonation with reformation of No. 15 occurred upon evaporation of the volatiles from CF_3CO_2H/CH_2Cl_2 solutions of No. 16 [31].

Decarbonylation of No. 15 in refluxing nonane for 3 h resulted in $(\mu\text{-H})_2Os_3(CO)_9(\mu_3\text{-}\eta^1\text{-}NC_6H_5)$ (Section 3.1.1.7.2.4) and $(\mu\text{-H})_2Os_3(\mu_3\text{-}\eta^2\text{-}NHC_6H_4)(CO)_9$ (see "Organoosmium Compounds" B 5, 1994, p. 285) [1]; the yield of both compounds decreased significantly at prolonged reaction times (up to 19 h) due to decomposition. $(\mu\text{-H})_2Os_3(CO)_9(\mu_3\text{-}\eta^1\text{-}NC_6H_5)$ was also obtained in 30% yield by pyrolysis of No. 15 in decahydronaphthalene at 196 to 198 °C for 24 h in a sealed tube [1].

$(\mu\text{-H})Os_3(CO)_{10}(\mu_2\text{-}\eta^1\text{-}NHC_6F_5)$ (Table 1, No. 19) crystallizes in the monoclinic space group $P2_1/c - C_{2h}^5$ (No. 14) with $a = 8.540(3)$, $b = 22.677(15)$, $c = 12.847(5)$ Å, $\beta = 116.70(2)°$; $Z = 4$. However, the interpretation of the structure established that a space group with the lower Pc symmetry is realized instead of the holohedral group. The crystal contains two crystallographically independent molecules, Molecule A (Fig. 2) and Molecule B (Fig. 3), which are identical in structure and composition. For both molecules, A and B, the Os(2)-μ-N-Os(3) planes form angles of 73° or 76° with the Os_3 planes, respectively. The planes of the C_6F_5 rings are oriented at 87° and 89° angles, and at 88° and 80° angles

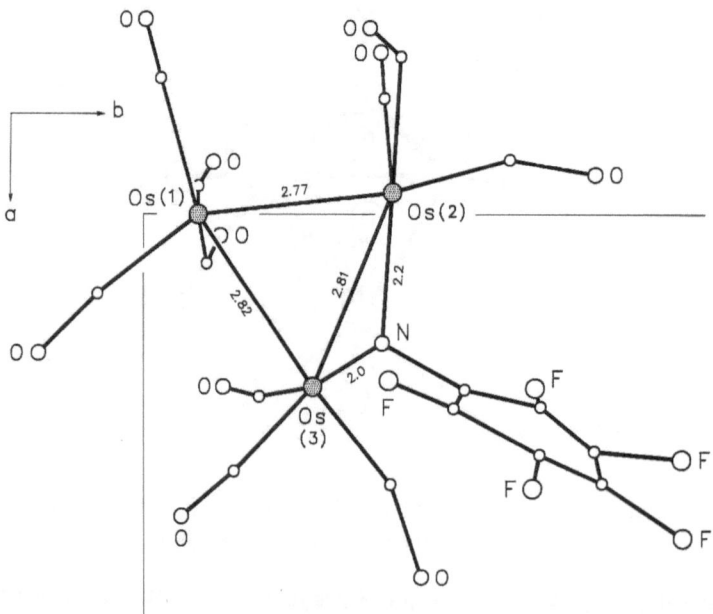

Fig. 2. Molecular structure of Molecule A of $(\mu\text{-H})Os_3(CO)_{10}(\mu_2\text{-}\eta^1\text{-}NHC_6F_5)$ (No. 19) with selected bond distances (in Å) [32].

References on pp. 24/5

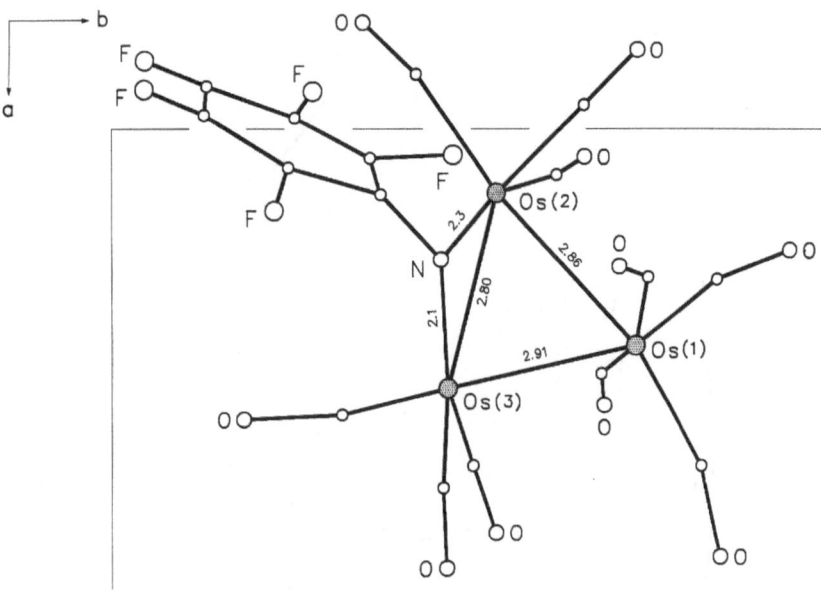

Fig. 3. Molecular structure of Molecule **B** of $(\mu\text{-H})Os_3(CO)_{10}(\mu_2\text{-}\eta^1\text{-NHC}_6F_5)$ (No. 19) with selected bond distances (in Å) [32].

to the Os(2)–μ–N–Os(3) and Os_3 triangles of Molecules **A** and **B**, respectively, being almost parallel to the bisecting planes of the Os_3N fragments. The bridging hydride ligands were not located by X-ray diffraction analysis, but are believed to span the Os(2)–Os(3) bond edges as the $\mu_2\text{-}\eta^1\text{-NHC}_6F_5$ ligand [32].

$(\mu\text{-H})Os_3(CO)_{10}(\mu_2\text{-}\eta^1\text{-NHSO}_2C_6H_4CH_3\text{-4})$ (Table 1, No. **23**) crystallizes in the triclinic space group $P\bar{1}\text{-}C_i^1$ (No. 2) with a = 8.2462(14), b = 8.6681(14), c = 18.5723(28) Å, α = 89.923(13)°, β = 98.614(13)°, γ = 113.046(12)°; Z = 2, D_c = 2.82 g/cm³. The structure is shown in **Fig. 4**. Both the $NHSO_2C_6H_4CH_3$ and the hydride ligand bridge the Os(1)–Os(2) bond edge; the Os(3)–Os(1)–N and Os(3)–Os(2)–N angles amounted to 83.3° on the average, the Os(1)–N–Os(2) angle is 81.6°. The $\mu_2\text{-}\eta^1\text{-NHSO}_2C_6H_4CH_3$ ligand occupies a symmetrical site in the isosceles Os_3 triangle with the $SO_2C_6H_4CH_3$ group in an exo configuration on the nitrogen atom, leaving the N–H system in an orientation to permit weak hydrogen bonding to the axial carbonyl group on the same side of the Os_3 plane on the Os(3) center. The plane of the $C_6H_4CH_3$ ring forms an angle of 52.67° with the Os_3 system. The $(\mu\text{-H})Os_3(CO)_{10}$-$(\mu_2\text{-}\eta^1\text{-N})$ group shows an approximate C_s symmetry and the substituents on the N and S atoms provide a staggered conformation about the N–S bond. The μ-hydride ligand was located directly and refined; the Os(1)–μ–H–Os(2) angle amounts to 93.5° [8].

$(\mu\text{-H})Os_3(CO)_{10}\{\mu_2\text{-}\eta^1\text{-NHS=N(Si(CH}_3)_3)_2\}$ (Table 1, No. **24**) crystallizes in the orthorhombic space group Pcmb (Pbcm) – D_{2h}^{11} (No. 57) with a = 7.418(2), b = 16.235(2), c = 23.970(4) Å; Z = 4, D_c = 2.44 g/cm³. The structure is shown in **Fig. 5**. The shortest metal–metal bond, the Os(2)–Os(2′) bond, is bridged by both the μ-hydride and the $\mu_2\text{-}\eta^1\text{-NHS=N(Si(CH}_3)_3)_2$ ligand. A mirror plane passing through the Os(1) center and the midpoint of the Os(2)–Os(2′) vector is oriented perpendicular to the Os_3 triangle. The angle defined by the Os(2)–N(1)–Os(3) plane and the Os_3 plane amounts to 73°. The N(1)–S and S–N(2) bonds of 1.78 and 1.69 Å, respectively, are both in the range of single bonds. The considerable

References on pp. 24/5

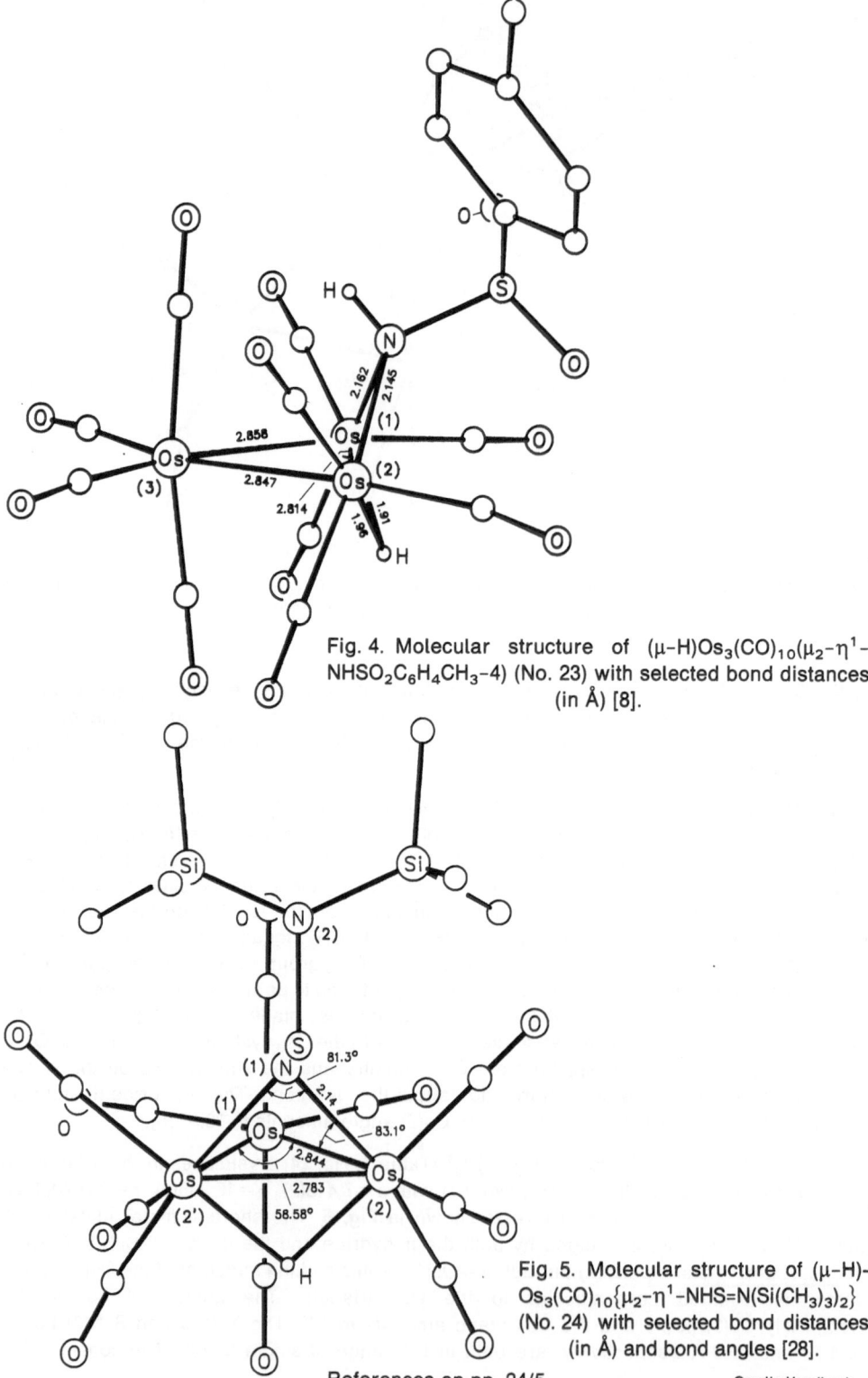

Fig. 4. Molecular structure of $(\mu\text{-H})Os_3(CO)_{10}(\mu_2\text{-}\eta^1\text{-}NHSO_2C_6H_4CH_3\text{-}4)$ (No. 23) with selected bond distances (in Å) [8].

Fig. 5. Molecular structure of $(\mu\text{-H})\text{-}Os_3(CO)_{10}\{\mu_2\text{-}\eta^1\text{-}NHS=N(Si(CH_3)_3)_2\}$ (No. 24) with selected bond distances (in Å) and bond angles [28].

References on pp. 24/5

differences of the bond lengths are partially explained as a result of different hybridizations of N(1) and N(2). N(1) exhibits a pyramidal configuration, whereas N(2) shows an essential trigonal-planar configuration. Both the μ-hydride and the hydrogen on N(1) were not observed crystallographically; the μ-H ligand is shown in a calculated position. The structure is discussed in detail [28].

$(μ-H)Os_3(CO)_{10}\{μ_2-η^1-NHN=C(CH_3)C_6H_5\}$ (Table 1, No. **27**) crystallizes in the monoclinic space group $P2_1/c - C_{2h}^5$ (No. 14) with a = 17.020(8), b = 9.181(3), c = 16.594(7) Å, β = 113.70(2)°; Z = 4, D_c = 2.75 g/cm³. The structure is shown in **Fig. 6**. The Os(1)–Os(2) bond edge is bridged by both the symmetrically coordinated $NHN=C(CH_3)C_6H_5$ moiety and the hydride ligand; the hydride was not observed cystallographically. The N(1)–N(2) bond length is slightly shorter than expected for a single bond, indicating some multiple-bond character, whereas the N(2)–C(1) bond distance corresponds to a localized double bond [14].

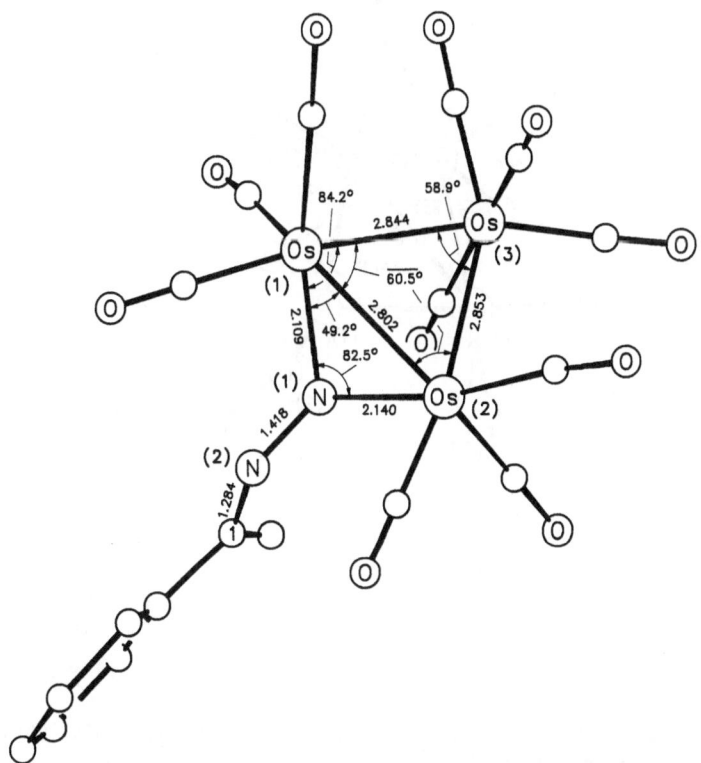

Fig. 6. Molecular structure of $(μ-H)Os_3(CO)_{10}\{μ_2-η^1-NHN=C(CH_3)C_6H_5\}$ (No. **27**) with selected bond distances (in Å) and bond angles [14].

$(μ-H)Os_3(CO)_{10}\{μ_2-η^1-NHN=C(C_6H_5)_2\}$ (Table 1, No. **28**) crystallizes in the monoclinic space group $P2_1/c - C_{2h}^5$ (No. 14) with a = 13.667, b = 12.015, c = 16.824 Å, β = 101.312°; Z = 4, D_c = 2.57 g/cm³ [10] or with a = 13.645(7), b = 11.948(8), c = 16.784(11) Å, β = 101.36(6)°; Z = 4, D_c = 2.59 g/cm³ [14]. The structure, shown in **Fig. 7**, is quite similar to that of No. 27, both corresponding to the isomers of type IIIa and IIIc (see p. 2) [14]. The crystal consists of discrete molecular units of $(μ-H)Os_3(CO)_{10}\{μ_2-η^1-NHN=C(C_6H_5)_2\}$. The arrangement of the carbonyl ligands provides approximate C_s symmetry to the basic $Os_3(CO)_{10}$ framework.

References on pp. 24/5

However it is suggested that this approximate C$_s$ symmetry extends outwards to include the NHN=C(C$_6$H$_5$)$_2$ ligand. Perfect mirror symmetry within the bridging ligand is disrupted by slight rotations about C(1)–C(2a) and C(1)–C(2b) bonds. Thus the dihedral angle between the planes defined by N(2)–C(1)–C(2a)–C(2b) and by the C(2b) phenyl ring is 8.98°. The Os(1)–N(1)–Os(2) plane makes an angle of 106.86° with the Os$_3$ plane. The bridging hydride was not located directly [10].

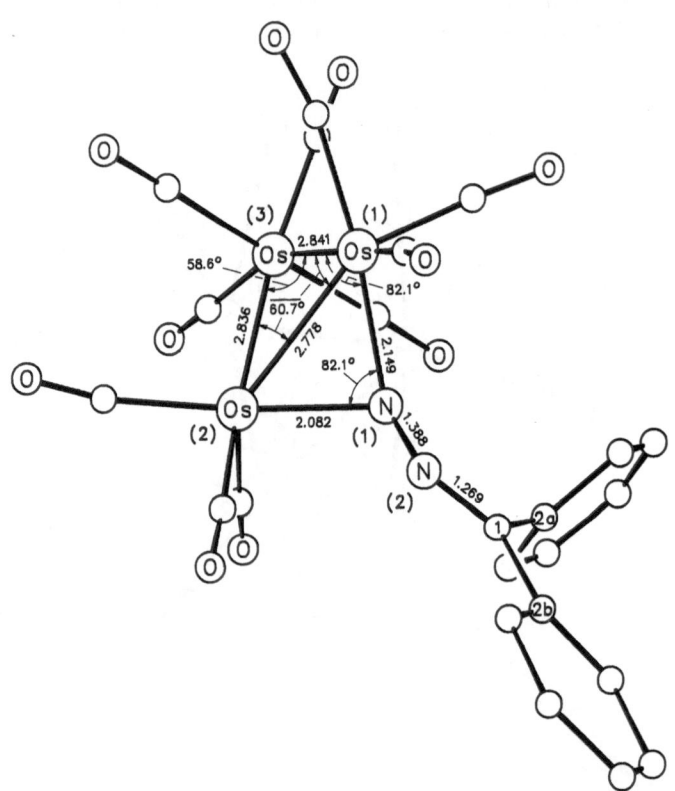

Fig. 7. Molecular structure of (μ-H)Os$_3$(CO)$_{10}${μ$_2$-η^1-NHN=C(C$_6$H$_5$)$_2$} (No. 28) [10, 14] with selected bond distances (in Å) and bond angles [14].

(μ-H)Os$_3$(CO)$_{10}${μ$_2$-η^1-NHSi(CH$_3$)$_3$} (Table 1, No. 30) crystallizes in the monoclinic space group P2$_1$/c – C$_{2h}^5$ (No. 14) with a = 17.557(5), b = 9.095(4), c = 13.503(5) Å, β = 90.65(2)°; Z = 4, D$_c$ = 2.90 g/cm^3. The structure is shown in **Fig. 8**. The shortest edge of the Os$_3$ triangle, the Os(1)–Os(2) bond, is bridged by the NHSi(CH$_3$)$_3$ ligand. Neither the N-bonded proton nor the hydride were located directly, but their presence was indicated by ^{1}H NMR spectroscopy. The μ-H ligand also bridges the Os(1)–Os(2) bond as concluded by the distribution of the CO groups around these two Os centers [18].

References on pp. 24/5

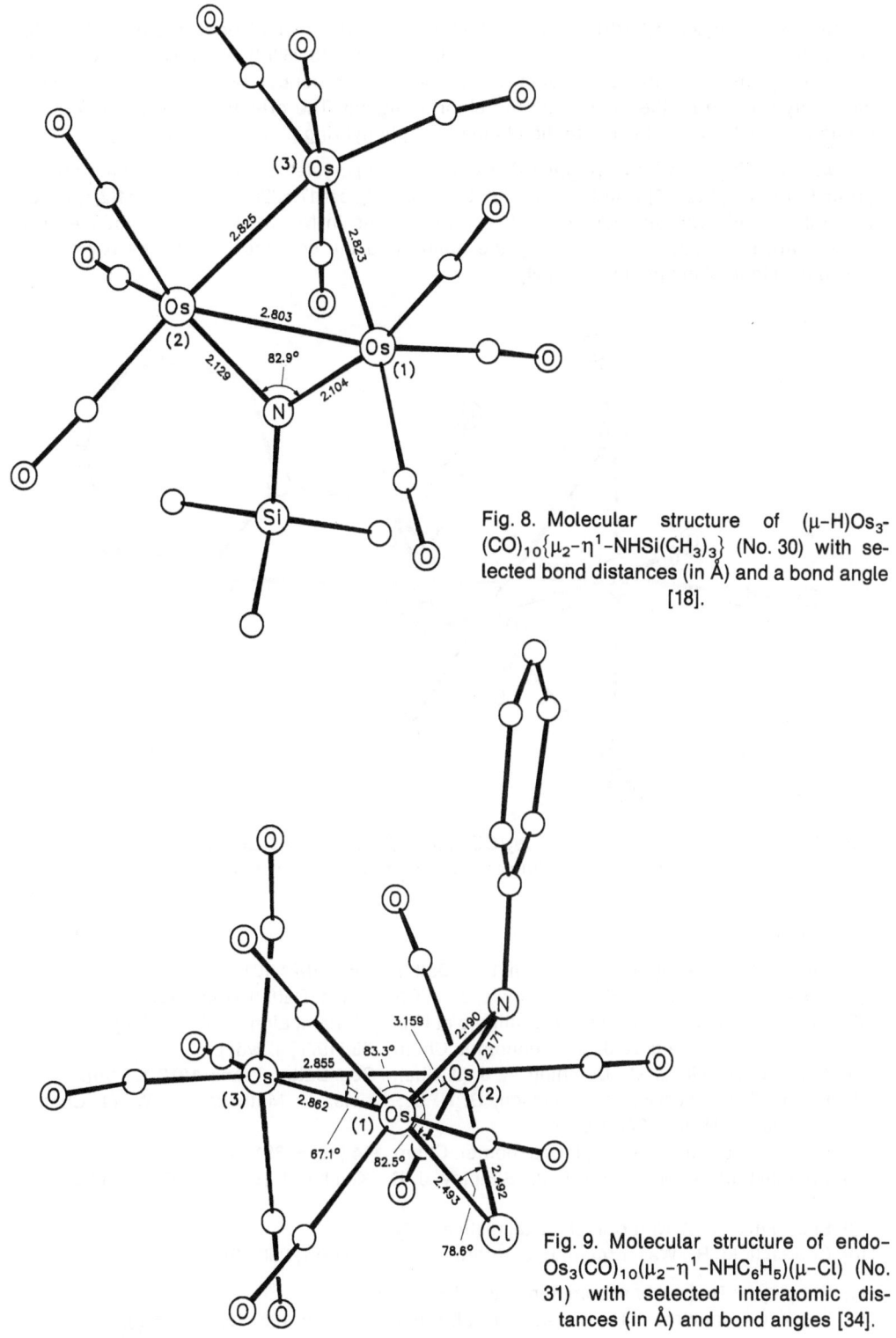

Fig. 8. Molecular structure of (μ–H)Os$_3$-(CO)$_{10}${μ$_2$-η^1-NHSi(CH$_3$)$_3$} (No. 30) with selected bond distances (in Å) and a bond angle [18].

Fig. 9. Molecular structure of endo-Os$_3$(CO)$_{10}$(μ$_2$-η^1-NHC$_6$H$_5$)(μ-Cl) (No. 31) with selected interatomic distances (in Å) and bond angles [34].

endo-Os₃(CO)₁₀(μ₂-η¹-NHC₆H₅)(μ-Cl) (Table 1, No. **31**) crystallizes in the orthorhombic space group Pccn − D$_{2h}^{10}$ (No. 56) with a = 26.423(5), b = 13.069(2), c = 12.446(2) Å; Z = 8, D$_c$ = 3.020 g/cm³. The structure is shown in **Fig. 9**. The two nonbonded Os centers are symmetrically bridged by the amino and the chloride ligand. The orientation of the N–H bond is approximately perpendicular to the plane of the phenyl ring [34].

exo-Os₃(CO)₁₀(μ₂-η¹-NHC₆H₅)(μ-Cl) (Table 1, No. **32**) crystallizes in the monoclinic space group C2/c − C$_{2h}^6$ (No. 15) with a = 16.286(3), b = 17.059(3), c = 17.422(4) Å, β = 118.51(2)°; Z = 8, D$_c$ = 3.052 g/cm³. The structure is shown in **Fig. 10**. As in No. 31, the two nonbonded Os centers are symmetrically bridged by the amino and the chloride ligand. The N–H bond lies in the plane of the phenyl ring [34].

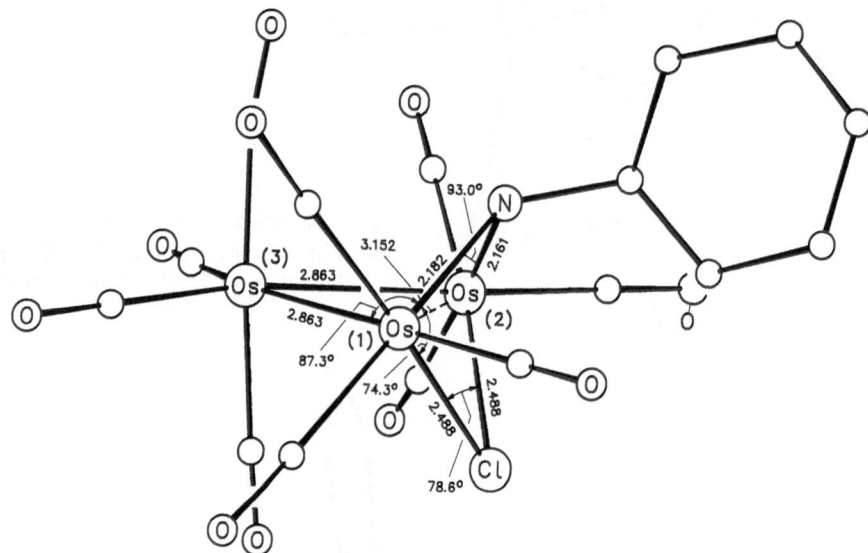

Fig. 10. Molecular structure of exo-Os₃(CO)₁₀(μ₂-η¹-NHC₆H₅)(μ-Cl) (No. 32) with selected interatomic distances (in Å) and bond angles [34].

References:

[1] Yin, C. C.; Deeming, A. J. (J. Chem. Soc. Dalton Trans. **1974** 1013/7).
[2] Bryan, E. G.; Johnson, B. F. G.; Lewis, J. (J. Chem. Soc. Dalton Trans. **1977** 1328/30).
[3] Keister, J. B. (Ph. D. Thesis Univ. Illinois at Urbana/Champaign 1977 from [10]).
[4] Yin, C. C.; Deeming, A. J. (J. Organomet. Chem. **133** [1977] 123/38).
[5] Azam, K. A.; Yin, C. C.; Deeming, A. J. (J. Chem. Soc. Dalton Trans. **1978** 1201/6).
[6] Bryan, E. G.; Forster, A.; Johnson, B. F. G.; Lewis, J.; Matheson, T. W. (J. Chem. Soc. Dalton Trans. **1978** 196/8).
[7] Yin, C. C.; Deeming, A. J. (J. Organomet. Chem. **144** [1978] 351/5).
[8] Churchill, M. R.; Hollander, F. J.; Shapley, J. R.; Keister, J. B. (Inorg. Chem. **19** [1980] 1272/77).
[9] Süss–Fink, G. (Z. Naturforsch. **35b** [1980] 454/7).
[10] Churchill, M. R.; Wasserman, H. J. (Inorg. Chem. **20** [1981] 2905/9).

[11] Shapley, J. R. (personal communication from [10]).
[12] Lin, Y. C.; Knobler, C. B.; Kaesz, H. D. (J. Am. Chem. Soc. **103** [1981] 1216/8).

[13] Banford, J.; Dawoodi, Z.; Henrick, K.; Mays, M. J. (J. Chem. Soc. Chem. Commun. **1982** 554/6).

[14] Burgess, K.; Johnson, B. F. G.; Lewis, J.; Raithby, P. R. (J. Chem. Soc. Dalton Trans. **1982** 263/9).

[15] Burgess, K.; Johnson, B. F. G.; Lewis, J.; Raithby, P. R. (J. Chem. Soc. Dalton Trans. **1982** 2085/92).

[16] Deeming, A. J.; Ghatak, I.; Owen, D. W.; Peters, R. (J. Chem. Soc. Chem. Commun. **1982** 392/3).

[17] Hursthouse, M. B.; Backer-Dirks, J. D. J.; Walker, N. (unpublished results from [16]).

[18] Johnson, B. F. G.; Lewis, J.; Raithby, P. R.; Sankey, S. W. (J. Organomet. Chem. **228** [1982] 135/8).

[19] Dawoodi, Z.; Mays, M. J.; Henrick, K. (J. Chem. Soc. Dalton Trans. **1984** 433/40).

[20] Maksakov, V. A.; Golubovskaya, É. V.; Korniets, E. D.; Chernii, I. V.; Gubin, S. P. (Izv. Akad. Nauk SSSR Ser. Kim. **1984** 1194/5; Bull. Acad. Sci. USSR Div. Chem. Sci. [Engl. Transl.] **33** [1984] 1097).

[21] Maksakov, V. A.; Gubin, S. P. (Koord. Khim. **10** [1984] 689/700; Sov. J. Coord. Chem. [Engl. Transl.] **10** [1984] 383/93).

[22] Banford, J.; Mays, M. J.; Raithby, P. R. (J. Chem. Soc. Dalton Trans. **1985** 1355/60).

[23] Lausarot, P. M.; Vaglio, G. A.; Valle, M.; Tiripicchio, A.; Tiripicchio Camellini, M.; Gariboldi, P. (J. Organomet. Chem. **291** [1985] 221/9).

[24] Maksakov, V. A.; Ershova, V. A.; Bragina, I. V. (Izv. Akad. Nauk SSSR Ser. Khim. **1985** 2829/30; Bull. Acad. Sci. USSR Div. Chem. Sci. [Engl. Transl.] **34** [1985] 2627).

[25] Odiaka, T. I. (J. Organomet. Chem. **284** [1985] 95/9).

[26] Odiaka, T. I.; Raithby, P. R.; Johnson, B. F. G.; Lewis, J. (unpublished work from [25]).

[27] Smieja, J. A.; Gladfelter, W. L. (J. Organomet. Chem. **297** [1985] 349/59).

[28] Süss-Fink, G.; Bühlmeyer, W.; Herberhold, M.; Gieren, A.; Hübner, T. (J. Organomet. Chem. **280** [1985] 129/38).

[29] Ditzel, E. J.; Johnson, B. F. G.; Lewis, J. (J. Chem. Soc. Dalton Trans. **1987** 1289/91).

[30] Ditzel, E. J.; Johnson, B. F. G.; Lewis, J. (J. Chem. Soc. Dalton Trans. **1987** 1293/7).

[31] Korniets, E. D.; Kedrova, L. K.; Maksimov, N. G. (Izv. Akad. Nauk SSSR Ser. Khim. **1987** 2064/9; Bull. Acad. Sci. USSR Div. Chem. Sci. [Engl. Transl.] **36** [1987] 1914/9).

[32] Podberezskaya, N. V.; Maksakov, V. A.; Bragina, I. V.; Korniets, E. D.; Ipatov, E. N.; Semyannikov, P. P.; Gubin, S. P. (Koord. Khim. **14** [1988] 978/84; Sov. J. Coord. Chem. [Engl. Transl.] **14** [1988] 543/9).

[33] Kabir, S. E.; Day, M.; Irving, M.; McPillips, T.; Minassian, H.; Rosenberg, E.; Hardcastle, K. I. (Organometallics **10** [1991] 3997/4004).

[34] Ramage, D. L.; Geoffroy, G. L.; Rheingold, A. L.; Haggerty, B. S. (Organometallics **11** [1992] 1242/55).

3.1.1.7.2.2 Compounds with $\mu_2\text{-}\eta^1$-Bonded NO Ligands

The compounds dealt with in this section all consist of $Os_3(CO)_{10-n}(\mu_2\text{-}\eta^1\text{-NO})_m^2 D_n$ skeletons, where m is 1 or 2, D is a N donor molecule, and n is 0 or 1. The compounds are represented by the general Formulas I, II, and III.

Formula I: $(CO)_3Os$ bridged by N–O and N–O groups to Os—CO (or ^2D) with two CO; $Os(CO)_4$ below.

Formula II: $[N\{P(C_6H_5)_3\}_2]$ salt with Os framework bearing $CO_{d'}$, $CO_{f'}$, N–O, CO_d, CO_f, $CO_{c'}$, $CO_{e'}$, CO_e, Os, CO_b, CO_a, CO_c.

I   II

Formula III: $(CO)_3Os$ bridged by N–O and H to $Os(CO)_3$; $Os(CO)_4$ below.

III

For the compounds of the type $Os_3(CO)_{10-n}(\mu_2\text{-}\eta^1\text{-NO})_2^2 D_n$ (Formula I) having two bridging NO ligands the Os_3 framework can be considered as an open four-center system with the nitrosyl groups acting as three-electron donors and only little direct metal-metal interactions within the $Os\text{-}(\mu\text{-NO})_2\text{-Os}$ moiety, evaluated for $Os_3(CO)_9(\mu_2\text{-}\eta^1\text{-NO})_2N(CH_3)_3$ (see Fig. 11) by X-ray determinations [5]. However, the anion of $[N\{P(C_6H_5)_3\}_2][Os_3(CO)_{10}(\mu\text{-NO})]$ (see Fig. 12) shows metal-metal bonding for the dibridged Os-Os vector [9].

$Os_3(CO)_9(\mu_2\text{-}\eta^1\text{-NO})_2N(CH_3)_3$ (see Formula I and Fig. 11) was obtained from $Os_3(CO)_{10}$-$(\mu_2\text{-}\eta^1\text{-NO})_2$ and $(CH_3)_3NO$. The crystals, obtained in the form of brown elongated rectangular blocks, are slightly air-sensitive [5]; no solvent and reaction conditions are given.

^{1}H NMR (no medium given): 2.63 ppm (s, CH_3) [5].

IR ($CHCl_3$): 1443 m cm^{-1} (NO); (no medium given, probably also in $CHCl_3$): 1943 m, 1993 m, 2003 s, 2023 vs, 2062 s, 2075 w, 2103 m cm^{-1} (all CO) [5].

Mass spectrum: $[M]^+$ [5].

The compound crystallizes in the triclinic space group $P\bar{1}\text{-}C_i^1$ (No. 2) with a = 12.859(3), b = 10.549(2), c = 17.356(5) Å, α = 112.63(2)°, β = 99.52(2)°, γ = 93.62(2)°; Z = 4, D_c = 2.95 g/cm^3. The asymmetric unit contains two independent, but structurally similar molecules; one of these is shown in **Fig. 11**. Both nitrosyl groups bridge the Os(1)-Os(2) edge symmetrically; the Os-NO bonds all amount to 2.05(1) Å. The elongated Os-$N(CH_3)_3$ bond reflects the steric requirements of the trimethylamine group. Only little direct metal-metal interactions between Os(1)-Os(2) were observed, suggesting an open four-center Os-$(NO)_2$-Os system with the nitrosyl groups acting as three-electron donors [5].

References on p. 29

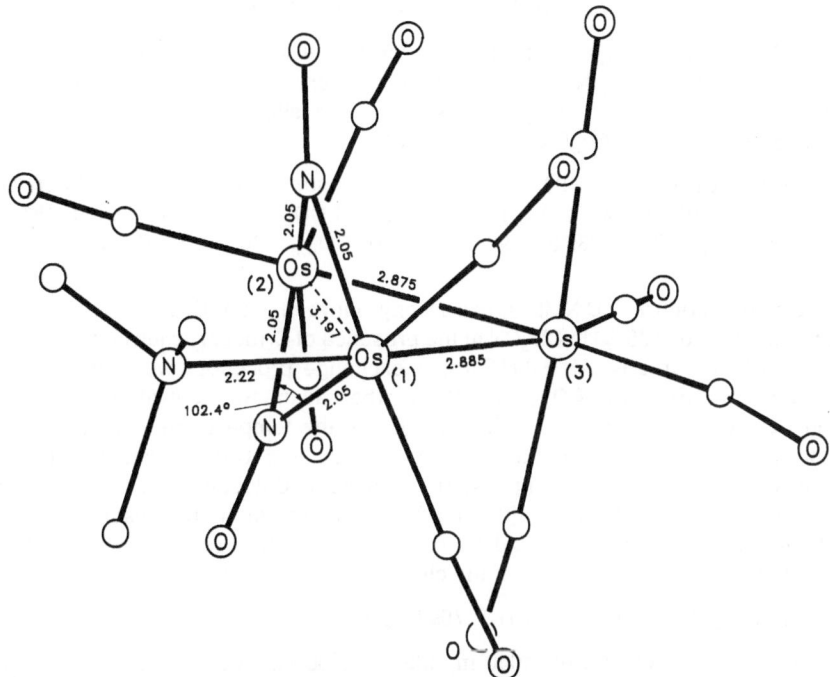

Fig. 11. Molecular structure of one of the two independent molecules of
$Os_3(CO)_9(\mu_2-\eta^1-NO)_2N(CH_3)_3$ with selected interatomic distances (in Å) and a bond angle
[5].

$Os_3(CO)_9(\mu_2-\eta^1-NO)_2NC_5H_5$ (Formula I) was prepared from $Os_3(CO)_9(NO)_2$ (see "Organoosmium Compounds" B 3, 1994, p. 247) and NC_5H_5 in refluxing hexane for 5 h; workup
by TLC with $CHCl_3$/cyclohexane as eluant gave green crystals in quantitative yield; a formation mechanism was discussed [4].

IR (cyclohexane): 1958 s, 1974 vs, 1994 m, 2006 vs, 2016 vs, 2050 m, 2061 vs, 2100 s,
2105 w cm^{-1}; absorptions due to NC_5H_5 occur in the same region as NO [4].

$Os_3(CO)_{10}(\mu_2-\eta^1-NO)_2$ (Formula I) was prepared by nitrosylation of $Os_3(CO)_{12}$ in benzene
under ca. 6.8 atm of NO at 100 °C for 24 h in a sealed tube, followed by crystallization
from heptane or by chromatography with hexane/CCl_4 (3:1) as eluant to yield air–stable,
light–green, platelike crystals [1, 2]. Only small amounts were obtained upon nitrosylation
in refluxing octane [3].

The title compound was also obtained by carbonylation of $Os_3(CO)_9(NO)_2$ (see "Organoosmium Compounds" B 3, 1994, p. 247) in cyclohexane at 81 °C for ca. 40 h (80%); a possible
formation mechanism was discussed [3, 4].

IR (KBr): 1484 s, 1503 m (both NO); (cyclohexane): 1996 m, 2008 s, 2014 w, 2025 s,
2054 s, 2063 s, 2068 s, 2108 w cm^{-1} (all CO) [3, 4]; similar data in C_2Cl_4 [1, 2].

Mass spectrum: $[M]^+$; a detailed fragmentation pattern examined by low–resolution mass
spectroscopy was given [1].

Decarbonylated in n–octane at 126 °C with formation of small amounts of $Os_3(CO)_9(NO)_2$
(see "Organoosmium Compounds" B 3, 1994, p. 247) [3].

                                    References on p. 29

Reaction with $(CH_3)_3NO$ yielded $Os_3(CO)_9(\mu_2-\eta^1-NO)_2N(CH_3)_3$ [5].

[N{P(C$_6$H$_5$)$_3$}$_2$][Os$_3$(CO)$_{10}$(μ_2-η^1-NO)] (see Formula II and Fig. 12) was prepared from $Os_3(CO)_{12}$ and $[N\{P(C_6H_5)_3\}_2]NO_2$ in THF at room temperature for 30 min, followed by precipitation of the product from hexane to give yellow–green, slightly air–sensitive microcrystals in 88% yield; intermediate $[Os_3(CO)_{10}NO]^-$ [7, 8].

^{13}C NMR (CH_2Cl_2/CD_2Cl_2, $-90\,°C$; for assignment, see Formula II): 178.3 (2 CO of $CO_{d,d'}$, $CO_{e,e'}$, or $CO_{f,f'}$), 179.0 (2 CO; $CO_{c,c'}$), 179.7 (2 CO of $CO_{d,d'}$, $CO_{e,e'}$, or $CO_{f,f'}$), 181.0, 186.6 (each 1 CO; $CO_{a,b}$; $J(C,C)=35.0$), 192.8 (2 CO of $CO_{d,d'}$, $CO_{e,e'}$, or $CO_{f,f'}$) ppm [9]; see also [10].

Variable–temperature ^{13}C NMR spectroscopy between $-90\,°C$ and $0\,°C$ in CD_2Cl_2 and between $+28\,°C$ and $+75\,°C$ in CD_3CN in the presence of $Cr(acac)_3$ showed fluxional behavior of the carbonyl groups. At $-30\,°C$ rapid exchange processes were observed for the three carbonyl groups $CO_{d,d'}$, $CO_{e,e'}$, and $CO_{f,f'}$, whereas the two equivalent carbonyl groups CO_c and $CO_{c'}$ exchange with $CO_{a,b}$ at $+28\,°C$. As the temperature was raised to $75\,°C$, two distinct processes average the carbonyl signals into two resonances at 183.4 and 185.8 ppm (spectra at various temperatures depicted). Calculations using the estimated $\Delta G^{\ddagger}_{298}$ value of ca. 21 kcal/mol (see [8]) and $\Delta\delta$ obtained from the two resonances indicated that a temperature of $+170\,°C$ would be required to observe coalescence of these resonances, but the decomposition point of the cluster is below that temperature [9].

^{15}N NMR (CH_2Cl_2; relative to $^{15}NH_3$): 759.8 ppm [8].

IR ($CHCl_3$): 1462 (NO); (THF): 1940 m, 1987 s, 2006 vs, 2014 s, 2073 w cm^{-1} (all CO) [8]; similar data in CH_2Cl_2 [7].

$[N\{P(C_6H_5)_3\}_2][Os_3(CO)_{10}(\mu_2-\eta^1-NO)]$ crystallizes in the monoclinic space group $P2_1/c-C^5_{2h}$ (No. 14) with $a=16.848(2)$, $b=17.015(2)$, $c=17.740(2)$ Å, $\beta=107.25(1)°$; $Z=4$, $D_c=1.94$ g/cm^3. The anion has approximately C_s symmetry with a triangular arrangement of the osmium atoms, see **Fig. 12**. The Os(1)–Os(2)–N plane forms a dihedral angle of $103.6°$ with the Os_3 plane. The nitrosyl oxygen is situated significantly out of the Os(1)–Os(2)–N plane, on the distal side with respect to Os(3). The NO ligand occupies a nearly axial position [9].

Treatment with CF_3SO_3H gave yellow crystals of $(\mu-H)Os_3(CO)_{10}(\mu_2-\eta^1-NO)$ [7].

(μ–H)Os$_3$(CO)$_{10}$(μ_2-η^1-NO) (Formula III) was obtained by protonation of $[N\{P(C_6H_5)_3\}_2]$-$[Os_3(CO)_{10}(\mu_2-\eta^1-NO)]$ with CF_3SO_3H [7] in CH_2Cl_2 at room temperature to give deep yellow crystals in 34% yield [8].

The title compound was also obtained from $Na[(\mu-H)Os_3(\mu-CO)(CO)_{10}]$ ("Organoosmium Compounds" B 3, 1994, pp. 55/64) and $[NO]PF_6$ in CH_3CN at room temperature for 1 h, followed by TLC with CH_2Cl_2/cyclohexane (1:10) as eluant [6].

IR (cyclohexane): 1586 (NO); 1987 m, 2008 s, 2020 m, 2029 vs, 2065 s, 2072 vs, 2113 w cm^{-1} (all CO) [6].

Mass spectrum: $[M]^+$ [6].

Hydrogenation of the title compound in heptane at $140\,°C$ for 2 h under 136 atm of H_2 gave the following four products by chromatography with hexane as eluant: $(\mu-H)Os_3$-$(CO)_{10}(\mu_2-\eta^1-NH_2)$ (26%; Section 3.1.1.7.2.1), $(\mu-H)_4Os_4(CO)_{12}$ (26%; see "Organoosmium Compounds" B 8, 1995, Section 4.1.1.3.1), $(\mu-H)_3HOs_3(CO)_8(\mu_3-\eta^1-NH)$ (30%), and $(\mu-H)_2Os_3(CO)_9(\mu_3-\eta^1-NH)$ (10%; both Section 3.1.1.7.2.4). The reaction was monitored by HPLC and/or IR spectroscopy; the HPLC chromatograms in CH_2Cl_2/hexane (9:1) at $140\,°C$ at various retention times were depicted [11].

Reaction with P(OCH$_3$)$_3$ in cyclohexane at room temperature for 1.5 h yielded (μ–H)Os$_3$-(CO)$_7${P(OCH$_3$)$_3$}$_3$(μ–NO) (Section 3.1.1.8.2) [6].

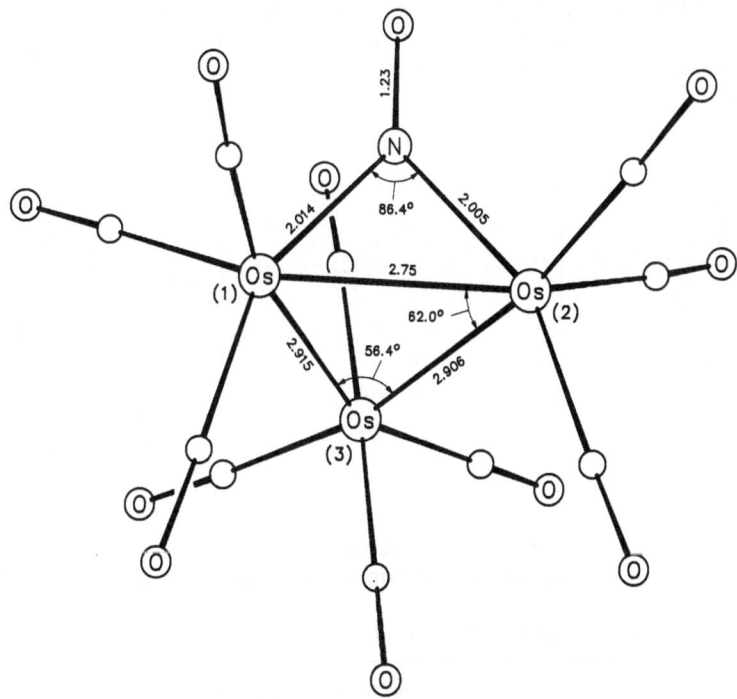

Fig. 12. Molecular structure of the anion of [N{P(C$_6$H$_5$)$_3$}$_2$][Os$_3$(CO)$_{10}$(μ$_2$-η^1-NO)] with selected bond distances (in Å) and bond angles [9].

References:

[1] Norton, J. R.; Collman, J. P.; Dolcetti, G.; Robinson, W. T. (Inorg. Chem. **11** [1972] 382/8).

[2] Norton, J. R.; Dolcetti, G. (Inorg. Syn. **16** [1976] 39/41).

[3] Bhaduri, S.; Johnson, B. F. G.; Lewis, J.; Watson, D. J.; Zuccaro, C. (J. Chem. Soc. Chem. Commun. **1977** 477/8).

[4] Bhaduri, S.; Johnson, B. F. G.; Lewis, J.; Watson, D. J.; Zuccaro, C. (J. Chem. Soc. Dalton Trans. **1979** 557/61).

[5] Johnson, B. F. G.; Lewis, J.; Raithby, P. R.; Zuccaro, C. (J. Chem. Soc. Chem. Commun. **1979** 916/7).

[6] Johnson, B. F. G.; Raithby, P. R.; Zuccaro, C. (J. Chem. Soc. Dalton Trans. **1980** 99/104).

[7] Stevens, R. E.; Yanta, T. J.; Gladfelter, W. L. (J. Am. Chem. Soc. **103** [1981] 4981/2).

[8] Stevens, R. E.; Gladfelter, W. L. (Inorg. Chem. **22** [1983] 2034/42).

[9] Johnson, B. F. G.; Lewis, J.; Mace, J. M.; Raithby, P. R.; Stevens, R. E.; Gladfelter, W. L. (Inorg. Chem. **23** [1984] 1600/3).

[10] Johnson, B. F. G.; Lewis, J.; Mace, J. M.; Raithby, P. R.; Stevens, R. E.; Gladfelter, W. L. (Inorg. Chem. **23** [1984] 2728).

[11] Smieja, J. A.; Gladfelter, W. L. (J. Organomet. Chem. **297** [1985] 349/59).

3.1.1.7.2.3 Compounds with Other μ$_2$-η^1-Bonded N Ligands

The compounds dealt with in this section are of the types $(\mu\text{-H})Os_3(CO)_{10}(\mu_2\text{-}\eta^1\text{-}$ $N=CR^1R^2)$ (Formula I), $[(\mu\text{-H})H_{0+n}Os_3(CO)_{10-n}(\mu_2\text{-}\eta^1\text{-NCO})]^{0/-}$ (Formula II), and $(\mu\text{-H})Os_3\text{-}$ $(CO)_{10}(\mu_2\text{-}\eta^1\text{-N=NR})$ (Formula III). Only $(\mu\text{-H})Os_3(CO)_{10}\{\mu_2\text{-}\eta^1\text{-NCH}(CF_3)P(C_2H_5)_3\}$ (No. 15), $(\mu\text{-H})Os_3(CO)_{10}(\mu_2\text{-}\eta^1\text{-NC}_3H_6)$ (No. 16), and $(\mu\text{-H})Os_3(CO)_{10}(\mu_2\text{-}\eta^1\text{-NC}_4H_8)$ (No. 17) deviate from these structures. The $\mu_2\text{-}\eta^1\text{-N=CR}^1R^2$, $\mu_2\text{-}\eta^1\text{-NCO}$, and $\mu_2\text{-}\eta^1\text{-N=NR}$ ligands act as three-electron donors.

Table 2
Compounds with Other μ$_2$-η^1-Bonded N Ligands.
An asterisk preceding the compound number indicates further information at the end of the table, pp. 37/42.
Explanations, abbreviations, and units on p. X.

No. compound	method of preparation (yield) properties and remarks

compounds of the type $(\mu\text{-H})Os_3(CO)_{10}(\mu_2\text{-}\eta^1\text{-N=CR}^1R^2)$ (Formula I)

*1 $(\mu\text{-H})Os_3(CO)_{10}(\mu_2\text{-}\eta^1\text{-N=CHCF}_3)$	from $(\mu\text{-H})_2Os_3(CO)_{10}$ and CF_3CN gas [5, 7, 8] in hexane at 80 °C for 16 h in a sealed tube, followed by TLC with CH_2Cl_2/hexane (5:95) as eluant (69%), together with $(\mu\text{-H})Os_3\{\mu_2\text{-}\eta^2\text{-C(CF}_3)\text{=NH}\}(CO)_{10}$ (14%; see "Organoosmium Compounds" B 5, 1994, p. 192) [5, 8]; similar preparation, but at room temperature for 12 h, followed by chromatography on alumina, gave No. 1 in ca. 35% yield, along with 32% of $(\mu\text{-H})Os_3\{\mu_2\text{-}\eta^2\text{-C(CF}_3)\text{=NH}\}(CO)_{10}$ [2] yellow crystals (from hexane at −5 °C) [2, 7]; m.p. 172 to 174 °C [2] ^{1}H NMR (CD_2Cl_2): −14.83 (s, OsH), 7.75 (q, CH; J(H,F)=4.2) [5]; $(CDCl_3)$: −14.84 (m, OsH), 7.74 (dq; ^{4}J(H,H?)=1.0, ^{3}J(H,F?)=4.5) [2] ^{19}F {^{1}H} NMR $(CD_2Cl_2$, vs. internal CCl_3F): 64.98 (dd; ^{3}J(F,CH)=4.5, ^{5}J(F,OsH)=0.7), 70.0 (s) [5]

References on pp. 42/3

Table 2 (continued)

No. compound	method of preparation (yield) properties and remarks
	IR (Nujol mull): 1140 vs, 1156 vs, 1270 s, 1370 m (all CF); (hexane): 1967 w, 1986 m, 2001 s, 2011 s, 2023 s, 2059 s, 2071 s, 2109 m (all CO) [5]; similar data in [2] mass spectrum: $[M]^+$ [5]
2 $(\mu\text{-H})Os_3(CO)_{10}(\mu_2\text{-}\eta^1\text{-N=CHCN})$	from $(\mu\text{-H})_2Os_3(CO)_{10}$ and $(CN)_2$ gas in CH_2Cl_2 at $-40\,°C$, then at room temperature for 3 d in a sealed tube (55%); together with 20% of isomeric $(\mu\text{-H})Os_3\{\mu_2\text{-}\eta^2\text{-C(CN)=NH}\}$-$(CO)_{10}$ (see "Organoosmium Compounds" B 5, 1994, p. 193) [18] yellow solid (from light petroleum/CH_2Cl_2) [18] ^1H NMR ($CDCl_3$, 27 °C): -14.90 (d, OsH), 7.59 (d, CH; J(μ-H,CH)=0.9) [18] IR (cyclohexane): 1987 m, 2003 s, 2011 s, 2024 vs, 2058 s, 2072 vs, 2107 m (all CO); (Nujol mull): 2220 (CN) [18]
*3 $(\mu\text{-H})Os_3(CO)_{10}(\mu_2\text{-}\eta^1\text{-N=CHC}_2H_5)$	by thermolysis of No. 16 in refluxing octane for 20 h and workup by TLC with CH_2Cl_2/hexane (1:9) as eluant (27%), along with $(\mu\text{-H})Os_3(\mu_3\text{-}\eta^2\text{-C}_3H_4NC\text{-cyclo})(CO)_9$ (15%), and $(\mu\text{-H})_3Os_5(\mu_2\text{-}\eta^2\text{-NC}_3H_4)(CO)_{14}$ (see "Organoosmium Compounds" B 8, 1995, Section 5.7; 3%) [20] yellow crystals (from CH_3OH by slow evaporation at 0 °C) [20] ^1H NMR (C_6D_6): -15.23 (s, OsH), 0.52 (t, CH_3; $^3J=7.7$), 1.90, 2.04 (ddq's, each 1 H of CH_2; $^2J=15.3$, $^3J=7.7$, 5.3 or 4.4), 6.45 (dd, CH; $^3J=5.3$ and 4.4) [20] IR (hexane): 1981 w, 1996 vs, 2003 vw, 2010 s, 2021 vs, 2055 s, 2067 vs, 2105 w (all CO) [20]
4 $(\mu\text{-H})Os_3(CO)_{10}(\mu_2\text{-}\eta^1\text{-N=CHC}_6H_5)$	from $Os_3(CO)_{12}$ and an excess of C_6H_5CN in refluxing n-octane for 6 h under H_2 in the presence of an excess of CH_3CO_2H; purified by TLC with petroleum ether/ether (19:1) as eluant (small amounts), along with $(\mu\text{-H})Os_3(CO)_{10}(\mu_2\text{-}\eta^1\text{-NHCH}_2C_6H_5)$ (Section 3.1.1.7.2.1) and $(\mu\text{-H})Os_3(CO)_{10}(\mu_2\text{-}\eta^2\text{-}O_2CCH_3)$ (see "Organoosmium Compounds" B 3, 1994, p. 124) as the main products; similar yields but considerably less $(\mu\text{-H})Os_3(CO)_{10}(\mu_2\text{-}\eta^1\text{-NHCH}_2C_6H_5)$ in the absence of CH_3CO_2H and/or H_2 or at shorter

Table 2 (continued)

No. compound	method of preparation (yield) properties and remarks

4 (continued)

reaction times; similar results were obtained starting from $(\mu-H)_2Os_3(CO)_{10}$ instead of $Os_3(CO)_{12}$ [16]

^{1}H NMR (CDCl$_3$): −14.68 (d, OsH), 7.34 to 7.62 (m, C$_6$H$_5$), 8.38 (d, CH); J(μ-H,CH)=0.9 [16]

IR (hexane): 1982 w, 1997 s, 2012 s, 2021 vs, 2056 s, 2067 vs, 2105 w (all CO) [16]

*5 $(\mu-H)Os_3(CO)_{10}\{\mu_2-\eta^1-N=CHN(CH_3)_2\}$

from $(\mu-H)_2Os_3(CO)_{10}$ and $(CH_3)_2NCN$ in refluxing hexane for 8 min, followed by extraction of the residue with ether for several times; further purification by TLC with CH$_2$Cl$_2$/ether (1:9) as eluant (76%); the reaction proceeded via intermediate $(\mu-H)Os_3(CO)_{10}NCN(CH_3)_2$ (see ''Organo-osmium Compounds'' B 3, 1994, p. 253) [15]

yellow platelets (from hot hexane at room temperature) [15]

^{1}H NMR (CD$_2$Cl$_2$): −14.52 (s, OsH), 3.07 (s, N(CH$_3$)$_2$), 7.08 (s, CH) [15]

IR (cyclohexane): 1974 m, 1982 sh, 1988 s, 1996 m, 2007 s, 2017 vs, 2051 s, 2061 s, 2101 m [15]

6 $(\mu-H)Os_3(CO)_{10}\{\mu_2-\eta^1-N=C(CH_3)_2\}$

from $Os_3(CO)_{11}NH=C(CH_3)_2$ (see ''Organo-osmium Compounds'' B 3, 1994, p. 256) in refluxing octane for ca. 8 h, followed by TLC with cyclohexane/CH$_2$Cl$_2$ (3:2) as eluant (16%) [11]

yellow, air-stable crystals (from CH$_2$Cl$_2$/pentane) [11]

^{1}H NMR (CDCl$_3$): −15.00 (s, OsH), 2.44 (s, (CH$_3$)$_2$); the appearance of only one singlet for the two CH$_3$ groups is indicative of the C_s symmetry of the molecule [11]

IR (KBr): 1655 (N=C); (pentane): 1979 w, 1993 vs, 2009 s, 2022 vs, 2054 s, 2065 vs, 2102 w (all CO) [11]

mass spectrum: [M]$^+$, [M−n CO]$^+$, n=1 to 10 [11]

decomposed between 170 and 200 °C [11]

7 $(\mu-H)Os_3(CO)_{10}(\mu_2-\eta^1-N=C_6H_{10}-cyclo)$

by thermolysis of $Os_3(CO)_{11}NH=C_6H_{10}-cyclo$ (see ''Organoosmium Compounds'' B 3, 1994, p. 256) in refluxing hydrocarbons at 120

References on pp. 42/3

Table 2 (continued)

No. compound	method of preparation (yield) properties and remarks
	to 130 °C for several hours (34 to 38%) [1, 11, 12]
	yellow crystals (from pentane) [1, 12]; decomposed between 170 and 200 °C [11]
	^1H NMR (CDCl$_3$): -15.00 (s, OsH), 1.67, 2.67 (m's, both CH$_2$) [1, 11]
	IR (KBr): 1645 (N=C); (pentane): 1984 w, 1998 vs, 2008 s, 2020 vs, 2055 s, 2063 vs, 2104 w (all CO) [1, 11]
	mass spectrum: [M]$^+$ [1], [M]$^+$, [M$-$n CO]$^+$, n$=$1 to 10 [11]
8 $(\mu\text{-H})Os_3(CO)_{10}\{\mu_2\text{-}\eta^1\text{-N=C}(CH_3)C_6H_5\}$	from $(\mu\text{-H})Os_3(CO)_{10}\{\mu_2\text{-}\eta^2\text{-}$ NHN=NC(=CH$_2$)C$_6$H$_5\}$ (Section 3.1.1.7.3.2) in 1,2-dimethoxyethane at 85 °C for 24 h (35%); a reaction mechanism was discussed [9]
	^1H NMR (CD$_2$Cl$_2$): -14.97 (s, OsH), 2.82 (s, CH$_3$), 7.10 to 7.50 (m, C$_6$H$_5$) [9]
	IR (hexane): 1978 w, 1994 s, 2006 m, 2020 s, 2054 s, 2068 s, 2104 w (all CO) [9]
	mass spectrum: [M]$^+$ [9]

compounds of the type $[(\mu\text{-H})H_{0+n}Os_3(CO)_{10-n}(\mu_2\text{-}\eta^1\text{-NCO})]^{0/-}$ (Formula II)

9 $[N\{P(C_6H_5)_3\}_2][(\mu\text{-H})HOs_3(CO)_9(\mu_2\text{-}\eta^1\text{-NCO})]$	by thermolysis of $[N\{P(C_6H_5)_3\}_2]$- $[(\mu\text{-H})HOs_3(CO)_{10}NCO]$ (see "Organoosmium Compounds" B 3, 1994, p. 252) in refluxing THF; extremely labile under CO atmosphere with reformation of the starting material [17]
	obtained as an intermediate in the preparation of $(\mu\text{-H})Os_3(C_4H_3O_3\text{-cyclo})(CO)_9(\mu_2\text{-}\eta^1\text{-NCO})$ (see "Organoosmium Compounds" B 5, 1994, p. 12) from $[N\{P(C_6H_5)_3\}_2]$- $[(\mu\text{-H})HOs_3(CO)_{10}NCO]$ with maleic anhydride [17]
	carbonylation leads to rapid reformation of $[N\{P(C_6H_5)_3\}_2][(\mu\text{-H})HOs_3(CO)_{10}NCO]$ [17]
10 $(\mu\text{-H})Os_3(CO)_{10}(\mu_2\text{-}\eta^1\text{-NCO})$	from $Os_3(CO)_{10}(NCCH_3)_2$ (see "Organoosmium Compounds" B 3, 1994, p. 218) and an excess of dry HNCO (in situ from (HNCO)$_3$ at 700 °C) in CH$_2$Cl$_2$ at room temperature (87%) [10]
	as a minor product from $Os_3(CO)_{10}(NCCH_3)_2$ with N$_2$H$_4$, based on the molecular ion in the mass spectrum [10]

References on pp. 42/3

Table 2 (continued)

No. compound	method of preparation (yield) properties and remarks

| 10 (continued) | ^{1}H NMR (CDCl$_3$): -13.82 (s, OsH) [10]
IR (cyclohexane): 1990 w, 2012 m, 2022 sh, 2027 s, 2065 m, 2077 s, 2112 w (all CO), 2226 s (NCO) [10]
mass spectrum: [M]$^+$ [10]
decomposed upon TLC on silica gel, being susceptible to nucleophilic addition [10]
reaction with NH$_2$NR$_2$ (R=H or CH$_3$) neat or in CH$_2$Cl$_2$ at room temperature led to (μ-H)Os$_3$(CO)$_{10}$(μ_2-η^1-NHCONHNR$_2$) (Section 3.1.1.7.2.1); similarly, treatment with ROH gave compounds of the type (μ-H)Os$_3$(CO)$_{10}$-(μ_2-η^1-NHCO$_2$R) [10] |
| 11 [N{P(C$_6$H$_5$)$_3$}$_2$][Os$_3$(CO)$_{10}$(μ_2-η^1-NCO)] | by thermolysis of [N{P(C$_6$H$_5$)$_3$}$_2$]-[Os$_3$(CO)$_{11}$NCO] (see "Organoosmium Compounds" B 3, 1994, p. 255) in refluxing THF for 3.5 h; not isolated, but identified on the base of the IR spectrum by comparison with that of [N{P(C$_6$H$_5$)$_3$}$_2$][Os$_3$(CO)$_{10}$(μ_2-η^1-NO)] (Section 3.1.1.7.2.2) [17]
IR (no medium given): 2217 (NCO); no other absorptions given [17] |

compounds of the type (μ-H)Os$_3$(CO)$_{10}$(μ_2-η^1-N=NR) (Formula III)

| 12 (μ-H)Os$_3$(CO)$_{10}$(μ_2-η^1-N=NC$_6$H$_5$) | from (μ-H)$_2$Os$_3$(CO)$_{10}$ and an excess of [C$_6$H$_5$N$_2$]BF$_4$ in refluxing CH$_2$Cl$_2$ for 10 h, followed by precipitation of [NH$_4$]BF$_4$ by bubbling dry NH$_3$ through the reaction mixture and TLC purification with petroleum ether as eluant (70%) [14]
by pyrolysis of (μ-H)Os$_3$(CO)$_{10}$(μ_2-η^2-N=NC$_6$H$_5$) (Section 3.1.1.7.3.2) in refluxing n-heptane (impure, not isolated) [14]
orange, air-stable crystals (from CH$_2$Cl$_2$/pentane); also stable in solution; soluble in common organic solvents [14]
^{1}H NMR (CDCl$_3$): -14.46 (s, OsH), 6.8 to 7.5 (m, C$_6$H$_5$) [14]
IR (cyclohexane): 1525 (N=N), 1955 w, 1980 w, 2003 vs, 2015 s, 2024 vs, 2059 s, 2069 vs, 2106 m (all CO); the ν(N=N) stretching frequency depended more strongly on the coordination mode of the N=NR ligand than |

References on pp. 42/3

Table 2 (continued)

No. compound	method of preparation (yield) properties and remarks
	on the electronic character of the aryl group R and the substituents; $R=C_6H_5$ (title compound), C_6H_4F-4 (No. 13), $C_6H_4CH_3-4$ (No. 14) [14] EI mass spectrum: $[M]^+$, $[M-n\,CO]^+$, $n=1$ to 10 [14] UV irradiation (450-W medium-pressure mercury-vapor lamp) in n-heptane for 4 to 6 h under 1 atm of CO yielded $(\mu-H)Os_3(CO)_{10}(\mu_2-\eta^2-N=NC_6H_5)$ (Section 3.1.1.7.3.2) [14] did not react with CO, H_2, gaseous HCl, or $P(C_6H_5)_3$ under mild conditions [14]
13 $Os_3(CO)_{10}(\mu_2-\eta^1-N=NC_6H_4F-4)$	from $(\mu-H)_2Os_3(CO)_{10}$ and an excess of $[4-FC_6H_4N_2]BF_4$ as described for No. 12 (59.1%) [14] by pyrolysis of $(\mu-H)Os_3(CO)_{10}(\mu_2-\eta^2-N=NC_6H_4F-4)$ (Section 3.1.1.7.3.2) in refluxing n-heptane (impure, not isolated) [14] orange, air-stable crystals (from CH_2Cl_2/pentane); also stable in solution; soluble in common organic solvents [14] ^1H NMR ($CDCl_3$): -14.43 (s, OsH), 6.45 to 6.91 (C_6H_4; J(H,H)$=9.3$; J(H,F)$=10.5$, 5.4) [14] ^{19}F (vs. internal C_6H_5F): 1.08 [14] IR (cyclohexane): 1525 (N=N), 1961 w, 1972 w, 1991 m, 2004 vs, 2012 s, 2025 vs, 2059 s, 2071 vs, 2108 m (all CO); for v(N=N), see also No. 12 [14] EI mass spectrum: $[M]^+$, $[M-n\,CO]^+$, $n=1$ to 10 [14] irradiation (450-W medium-pressure mercury-vapor lamp) in n-heptane for 4 to 6 h under 1 atm of CO yielded $(\mu-H)Os_3(CO)_{10}(\mu_2-\eta^2-N=NC_6H_4F-4)$ (Section 3.1.1.7.3.2) [14] did not react with CO, H_2, gaseous HCl, or $P(C_6H_5)_3$ under mild conditions [14]
*14 $(\mu-H)Os_3(CO)_{10}(\mu_2-\eta^1-N=NC_6H_4CH_3-4)$	from $(\mu-H)_2Os_3(CO)_{10}$ and an excess of $[4-CH_3C_6H_4N_2]BF_4$ as described for No. 12 (79.4%) [14] by pyrolysis of $(\mu-H)Os_3(CO)_{10}(\mu_2-\eta^2-N=NC_6H_4CH_3-4)$ (Section 3.1.1.7.3.2) in refluxing n-heptane (impure, not isolated) [14]

References on pp. 42/3

Table 2 (continued)

No. compound	method of preparation (yield) properties and remarks

*14 (continued)

orange, air-stable crystals (from CH$_2$Cl$_2$/pentane); also stable in solution; soluble in common organic solvents [14]

^{1}H NMR (CDCl$_3$): -14.36 (s, OsH), 2.44 (s, CH$_3$), 7.49, 7.55 (d's, each 2 H; J(H,H)=9.0) [14]

^{13}C $\{^1$H$\}$ NMR (CDCl$_3$): 164.6, 165.0, 168.4, 170.2, 171.5, 171.9, 175.2, 176.2, 176.7, 180.3 (each 1 CO); invariant between -55 °C and $+55$ °C, indicating that inversion of configuration at μ-N is slow on NMR time scale [14]

^{15}N NMR (CDCl$_3$?; vs. ^{15}NH$_3$): 448.6 (d; J(^{15}N,μ-H)=2) [14]

UV (n-heptane): 385; spectrum depicted [14]

IR (cyclohexane): 1525 (N=N), 1953 w, 1965 w, 1983 m, 2000 vs, 2013 s, 2021 vs, 2057 s, 2066 vs, 2106 m (all CO); for ν(N=N), see also No. 12 [14]

EI mass spectrum: [M]$^+$, [M$-$n CO]$^+$, n=1 to 10 [14]

other compounds

15 (μ-H)Os$_3$(CO)$_{10}\{$μ$_2$-η^1-NCH(CF$_3$)P(C$_2$H$_5$)$_3\}$

from No. 1 and a slight excesss of P(C$_2$H$_5$)$_3$ in refluxing hexane for 4 h, purified by TLC with hexane as eluant (80%) [5]

^{1}H NMR (CD$_2$Cl$_2$): -14.68 (d, OsH; J(H,P)=8.1), 7.88 (m, CH; the eight line coupling, expected to be caused by the P and F nuclei, is only partially resolved) [5]

^{19}F $\{^1$H$\}$ NMR (CD$_2$Cl$_2$, vs. internal CCl$_3$F): 70.61 (ddd; ^{3}J(F,P)=2.9, ^{3}J(F,CH)=4.8, ^{5}J(F,OsH)=0.7), 71.4 (d; ^{3}J(F,P)=3.0) [5]

IR (hexane): 1941 m, 1962 w, 1974 m, 1984 s, 1992 s, 2002 sd, 2006 s, 2014 s, 2055 vs, 2094 s [5]

mass spectrum: [M]$^+$ [5]

*16 (μ-H)Os$_3$(CO)$_{10}$(μ$_2$-η^1-NC$_3$H$_6$)

from Os$_3$(CO)$_{10}$(NCCH$_3$)$_2$ (see "Organoosmium Compounds" B 3, 1994, p. 218) and azetidine in refluxing CH$_2$Cl$_2$ for 6 h; workup by TLC with CH$_2$Cl$_2$/hexane (1:9) as eluant (34%) [20]

orange crystals (from CH$_3$CN by slow evaporation at 25 °C) [20]

References on pp. 42/3

Table 2 (continued)

No. compound	method of preparation (yield) properties and remarks
	^{1}H NMR (CDCl$_3$): -14.00 (s, OsH), 2.53 (q, 2 H), 3.99, 4.43 (t's, each 2 H); ^{3}J(H,H) $= 6.9$ [20]
	IR (hexane): 1976 w, 1986 s, 1997 vw, 2007 m, 2019 s, 2050 m, 2062 vs, 2103 w (all CO) [20]

17 (μ-H)Os$_3$(CO)$_{10}$(μ$_2$-η^1-NC$_4$H$_8$)

from Os$_3$(CO)$_{10}$(NCCH$_3$)$_2$ (see "Organoosmium Compounds" B 3, 1994, p. 218) and pyrrolidine in C$_6$H$_6$ at 40 to 45 °C (7%); together with 50% of (μ-H)Os$_3$(CO)$_{10}$(μ$_2$-η^2-NC$_4$H$_6$) (see "Organoosmium Compounds" B 5, 1994, p. 213) and 3% of (μ-H)Os$_3$(CO)$_{10}$-(μ$_2$-η^2-O=CNC$_4$H$_8$) (see "Organoosmium Compounds" B 5, 1994, p. 117) [19]

^{1}H NMR (CDCl$_3$): -13.69 (s, OsH), 1.84 (m, 4 H), 3.26, 3.84 (t's, each 2 H; J(H,H) $= 7.2$) [19]

IR (cyclohexane): 1972 w, 1982 s, 2002 s, 2015 vs, 2046 s, 2057 vs, 2099 w (all CO) [19]

thermolysis in refluxing heptane for 4 h led quantitatively to (μ-H)Os$_3$(CO)$_{10}$(μ$_2$-η^2-NC$_4$H$_6$) (see "Organoosmium Compounds" B 5, 1994, p. 213) [19]

*Further information:

(μ-H)Os$_3$(CO)$_{10}$(μ$_2$-η^1-N=CHCF$_3$) (Table 2, No. 1) crystallizes in the monoclinic space group P2$_1$/n (P2$_1$/c) $-$ C$_{2h}^5$ (No. 14) [6, 7] with a $= 7.321(5)$, b $= 29.21(1)$, c $= 9.364(5)$ Å, β $= 100.55(5)°$; Z $= 4$, D$_c$ $= 3.196$ g/cm^3 (X-ray diffraction) at ambient temperature? [7] and a $= 7.264(2)$, b $= 28.385(7)$, c $= 9.256(2)$ Å, β $= 99.52(2)°$; D$_c$ $= 3.345$ g/cm^3 (neutron diffraction) at 20 K [6]. The molecular structure at 20 K is shown in **Fig. 13**. The crystalline complex consists of isolated molecules separated by normal van der Waal's distances. The short Os(1)–Os(3) bond edge is symmetrically bridged by the hydride and the N=CHCF$_3$ ligand. The 48-electron cluster has three formal Os-Os bonds, and disregarding the Os(1)–Os(3) interaction, all metal centers have approximately octahedral environment, obeying the effective-atomic-number rule. The Os-N-Os and N-C-H-C planes are coplanar within 0.5 Å, indicating that the alkylideneimido ligand displays a near planarity at the N=C unit. These results are consistent with sp hybridization at N and sp^2 at C. The dihedral angles between the Os$_3$ and Os$_2$H and Os$_2$N planes are 61.4 and 72.1°, respectively. The N=CHCF$_3$ fragment is slightly tilted with respect to the perpendicular Os(1)–Os(3) vector by 142.2° for Os(1)-N-C, and by 132.3° for Os(3)-N-C, probably resulting from steric crowding caused by the bulky CF$_3$ group. An approximate mirror plane of symmetry passes perpendicularly the Os(1)–Os(3) vector and through Os(2) [6]. Similar structural features have been found by X-ray analysis [7].

References on pp. 42/3

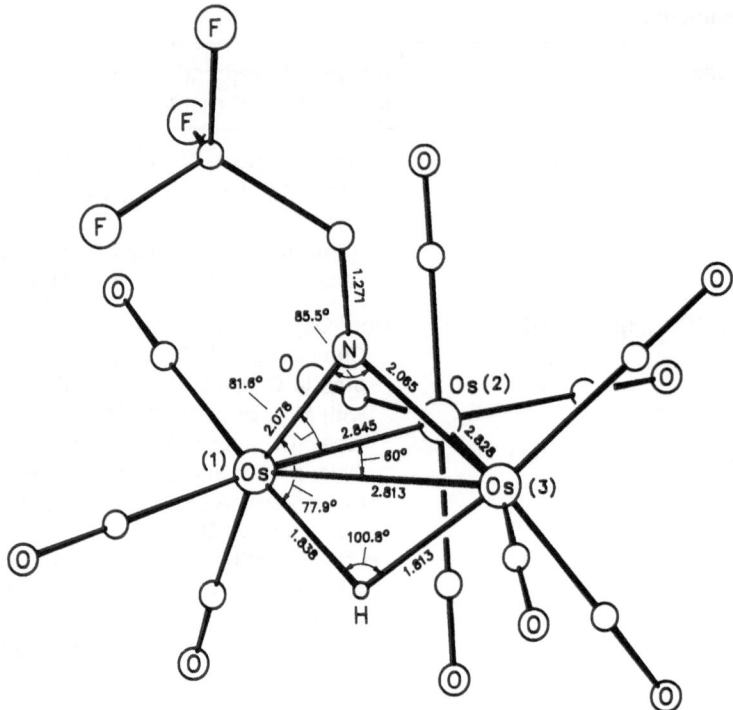

Fig. 13. Molecular structure of $(\mu\text{-H})Os_3(CO)_{10}(\mu_2\text{-}\eta^1\text{-N=CHCF}_3)$ (No. 1) with selected bond distances (in Å) and bond angles (obtained from neutron diffraction) [6].

Thermolysis above 150 °C neat [13] or in nonane for 20 h [5] afforded $(\mu\text{-H})Os_3(\mu_3\text{-}\eta^2\text{-}$ $N=CHCF_3)(CO)_9$ ("Organoosmium Compounds" B 5, 1994, p. 300) in low yields together with extensive decomposition to Os metal [5, 13]. Refluxing in octane caused no CO elimination, neither by cluster rearrangement nor by decomposition [13].

Hydrogenation with ca. 49 atm of H_2 in hexane at 140 °C for 16 h led to approximately equal amounts of $(\mu\text{-H})Os_3(CO)_{10}(\mu_2\text{-}\eta^1\text{-NHCH}_2CF_3)$ (Section 3.1.1.7.2.1), $(\mu\text{-H})_2Os_3(CO)_9\text{-}$ $(\mu_3\text{-}\eta^1\text{-NCH}_2CF_3)$, and $(\mu\text{-H})_3HOs_3(CO)_8(\mu_3\text{-}\eta^1\text{-NCH}_2CF_3)$ (both Section 3.1.1.7.2.4); a reaction mechanism was proposed and discussed [8, 13].

Reaction with $P(CH_3)_2C_6H_5$ in refluxing heptane gave a mixture of the syn and anti isomers of $(\mu\text{-H})Os_3(CO)_9P(CH_3)_2C_6H_5(\mu\text{-N=CHCF}_3)$ (see "Organoosmium Compounds" B 4b, in preparation) [2, 3]. Treatment of No. 1 with a slight excess of $P(C_2H_5)_3$ in refluxing hexane for 4 h yielded No. 15 [5].

$(\mu\text{-H})Os_3(CO)_{10}(\mu_2\text{-}\eta^1\text{-N=CHC}_2H_5)$ (Table 2, No. 3) crystallizes in the triclinic space group $P\bar{1}-C_i^1$ (No. 2) with a = 14.626(3), b = 22.961(6), c = 9.089(2) Å, $\alpha = 92.73(2)°$, $\beta = 99.03(2)°$, $\gamma =$ 81.52(2)°; Z = 6, $D_c = 3.03$ g/cm³. The crystal contains three independent but structurally similar molecules in the asymmetric unit; one of these (Molecule A) is shown in Fig. 14. The nitrogen of the $\mu_2\text{-}\eta^1\text{-N=CHC}_2H_5$ moiety bridges the Os(1)-Os(2) bond edge symmetrically. The hydride ligand spanning the same Os-Os edge was located and refined crystallographically [20].

References on pp. 42/3

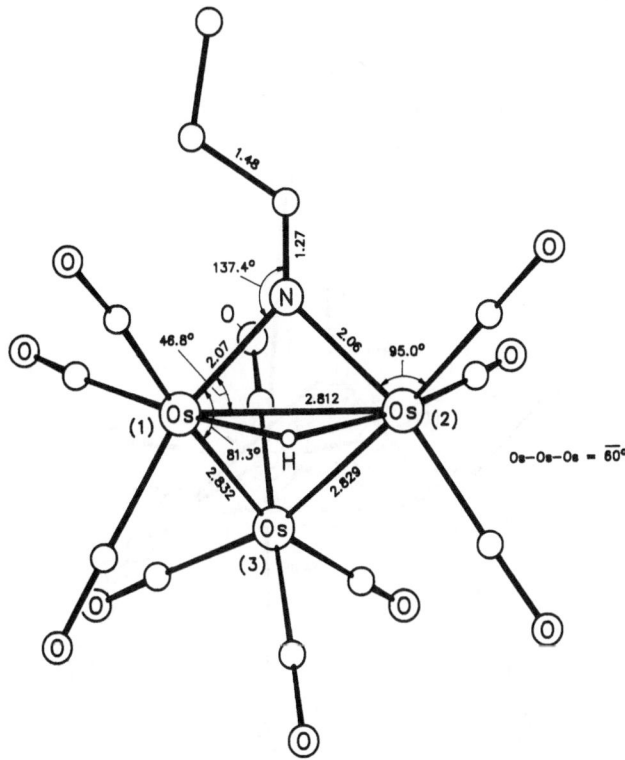

Fig. 14. Molecular structure of Molecule **A** of (μ-H)Os₃(CO)₁₀(μ₂-η¹-N=CHC₂H₅) (No. 3) with selected bond distances (in Å) and bond angles [20].

(μ-H)Os$_3$(CO)$_{10}${μ$_2$-η^1-N=CHN(CH$_3$)$_2$} (Table **2**, No. **5**) crystallizes in the triclinic space group P$\bar{1}$−C$_i^1$ (No. 2) with a = 7.403(1), b = 9.302(2), c = 15.067(4) Å, α = 94.75(2)°, β = 95.13(2)°, γ = 96.98(2)°; Z = 2, D$_c$ = 3.228 g/cm³. The structure is shown in **Fig. 15**. Both the μ-H and the N=CHN(CH$_3$)$_2$ ligands bridge the Os(1)-Os(2) bond edge on opposite sides of the Os$_3$ triangle; the hydride ligand was not located directly. The bridging N=CHN(CH$_3$)$_2$ ligand is approximately planar, suggesting some multiple bonding between N(1) and C(1), and between C(1) and N(2), involving the p orbitals on each of these atoms perpendicular to the plane of the ligand. The angle between the Os$_3$ triangle and the Os(1)-Os(2)-N(1) plane amounts to 108.5° [15].

Decarbonylation in refluxing decane for 10 h yielded (μ-H)Os$_3${μ$_3$-η^2-NH=CN(CH$_3$)$_2$}-(CO)$_9$; see "Organoosmium Compounds" B 5, 1994, p. 296 [15].

Reaction with HX gas (X=Cl) or HX/HBF$_4$·O(CH$_3$)$_2$ (X=OCH$_3$, OH, OCH$_2$CH$_2$OH) in CH$_2$Cl$_2$ at room temperature gave compounds of the type (μ-H)Os$_3$(CO)$_{10}$(μ-X) in good yields; in some cases (μ-H)Os$_3$(CO)$_{10}$(μ-OCH$_3$) was obtained as co-product; for the compounds, see "Organoosmium Compounds" B 3, 1994, pp. 94, 96, and 98). Similar treatment with CF$_3$CO$_2$H gave (μ-H)Os$_3$(CO)$_{10}$(μ$_2$-η^2-O$_2$CCF$_3$) (see "Organoosmium Compounds" B 3, 1994, p. 125) in 95% yield; the reaction proceeded via the N-protonated complex [(μ-H)Os$_3$(CO)$_{10}${μ$_2$-η^1-NHCH=N(CH$_3$)$_2$}][CF$_3$CO$_2$] (Section 3.1.1.7.2.1), which is only observed at −60 °C, based on ^{1}H NMR spectra. The reaction of No. 5 with the weak acid

                                        References on pp. 42/3

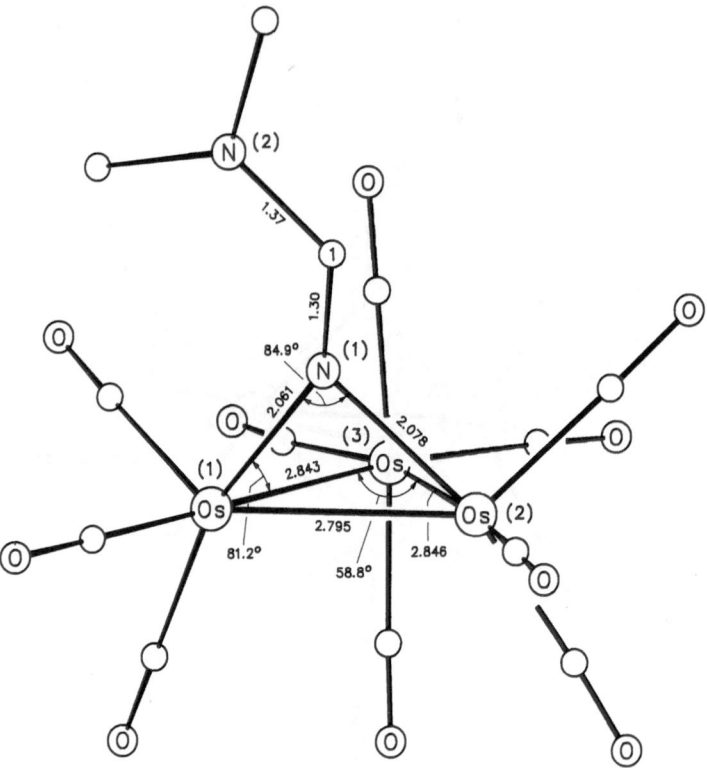

Fig. 15. Molecular structure of $(\mu\text{-H})Os_3(CO)_{10}\{\mu_2\text{-}\eta^1\text{-}N=CHN(CH_3)_2\}$ (No. 5) with selected bond distances (in Å) and bond angles [15].

CH_3CO_2H in the presence of $HBF_4 \cdot O(C_2H_5)_2$ led to $(\mu\text{-H})Os_3(CO)_{10}(\mu_2\text{-}\eta^2\text{-}O_2CCH_3)$ (see "Organoosmium Compounds" B 3, 1994, p. 124) in ca. 97%. Treatment with $HBF_4 \cdot O(C_2H_5)_2$ in CH_2Cl_2 resulted in a mixture of $(\mu\text{-H})Os_3(CO)_{10}(\mu\text{-OH})$ and $(\mu\text{-H})Os_3(CO)_{10}(\mu\text{-OC}_2H_5)$ (see "Organoosmium Compounds" B 3, 1994, pp. 94 and 98) in 21% and 58%, respectively [15].

Hydrogenation with 50 atm of H_2 at 140 °C for 18 h yielded $(\mu\text{-H})_4Os_4(CO)_{12}$ (see "Organoosmium Compounds" B 8, 1995, Section 4.1.1.3.1); attempted hydrogenation with 1 atm of H_2 at 69 °C for 12 h failed completely [15].

Treatment with $P(C_2H_5)_3$ in refluxing hexane gave low yields of an inseparable mixture of syn- and anti-$Os_3(CO)_9P(C_2H_5)_3\{\mu\text{-N}=CHN(CH_3)_2\}$ (see "Organoosmium Compounds" B 4b, in preparation) [15].

$(\mu\text{-H})Os_3(CO)_{10}(\mu_2\text{-}\eta^1\text{-}N=NC_6H_4CH_3\text{-4})$ (Table 2, No. 14) crystallizes in the monoclinic space group $P2_1/n$ $(P2_1/c) - C_{2h}^5$ (No. 14) with a = 9.401(2), b = 13.463(4), c = 18.110(6) Å, β = 99.85(2)°; Z = 4, D_c = 2.86 g/cm^3. The molecular structure is shown in **Fig. 16**. Both the μ-H and the $N=NC_6H_4CH_3\text{-4}$ ligands bridge the slightly shortened Os(1)–Os(3) bond edge; the Os(1)–H–Os(3) and Os(1)–N–Os(3) planes form an angle of 132.81°. The symmetrically coordinated hydride ligand is located crystallographically and refined. The angle between the Os_3 plane and the Os(1)–H–Os(3) or Os(1)–N–Os(3) planes amounts to 121.45° or 105.74°, respectively. The $(\mu\text{-H})Os_3(CO)_{10}$ part of the molecule has approximate C_s symmetry; how-

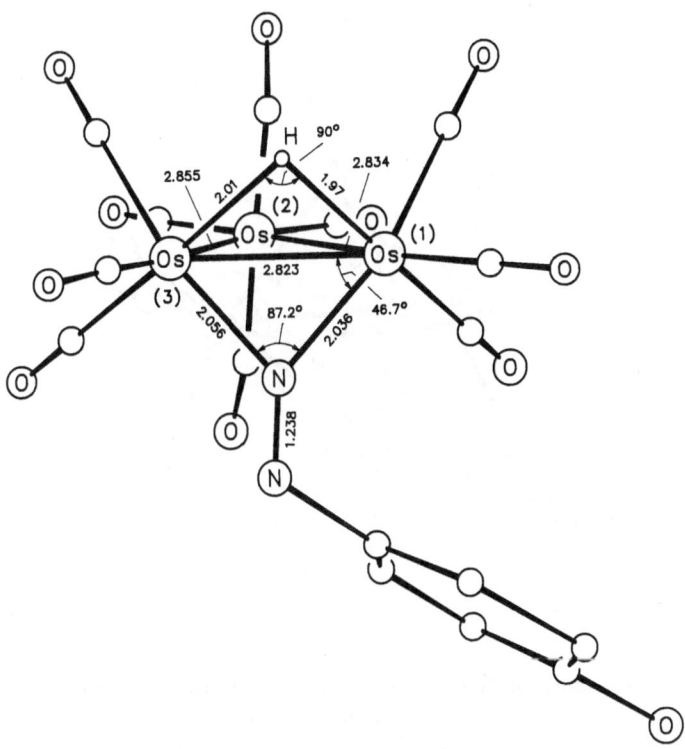

Fig. 16. Molecular structure of $(\mu\text{-H})Os_3(CO)_{10}(\mu_2\text{-}\eta^1\text{-N=NC}_6H_4CH_3\text{-4})$ (No. 14) with selected bond distances (in Å) and bond angles [4].

ever, this symmetry does not extend to include the bridging $N=NC_6H_4CH_3\text{-4}$ moiety. The orientation of the p-tolyl group destroys the potential C_s symmetry of the molecule since it lies essentially under the Os(1) center [4].

UV irradiation (450-W medium-pressure mercury-vapor lamp) in n-heptane for 4 to 6 h under 1 atm of CO resulted in $(\mu\text{-H})Os_3(CO)_{10}(\mu_2\text{-}\eta^2\text{-N=NC}_6H_4CH_3\text{-4})$ (Section 3.1.1.7.3.2) in 85% yield and 15% of unreacted No. 14. The quantum yields Φ for the photoisomerization at 313 and 366 nm (near the absorption maximum of 385 nm of No. 14) were estimated as 0.06 and 0.006, respectively, indicating that the photoisomerization resulted primarily from the absorption of the higher-energy photons and suggesting a transfer of the absorbed light energy from metal–metal σ orbitals to metal–ligand σ orbitals. Further irradiation led to slow decomposition but not to a further increase of the yield of $(\mu\text{-H})Os_3(CO)_{10}(\mu_2\text{-}\eta^2\text{-N=NC}_6H_4CH_3\text{-4})$ [14].

Treatment with CO, H_2, gaseous HCl, or $P(C_6H_5)_3$ under mild conditions caused no reactions [14].

$(\mu\text{-H})Os_3(CO)_{10}(\mu_2\text{-}\eta^1\text{-NC}_3H_6)$ (Table 2, No. 16) crystallizes in the monoclinic space group $P2_1/c - C_{2h}^5$ (No. 14) with a = 8.793(2), b = 16.265(2), c = 13.766(4) Å, β = 110.56°; Z = 4, 3.17 g/cm³; the structure is shown in Fig. 17. The azetidine ring is slightly puckered, and the methylene group at the 3-position was disordered between the two sites C′ and C″. Because of the disorder the C positions were not accurately determined leading to C–C

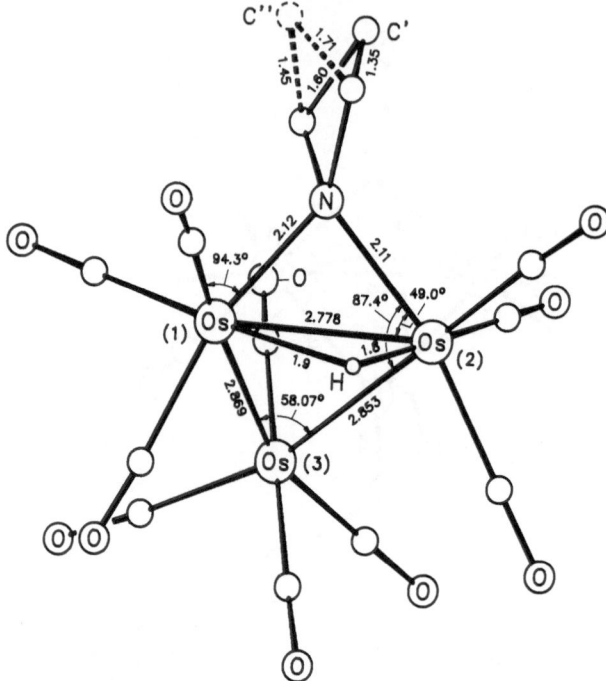

Fig. 17. Molecular structure of $(\mu\text{-H})Os_3(CO)_{10}(\mu_2\text{-}\eta^1\text{-}NC_3H_6)$ (No. 16) with selected bond distances (in Å) and bond angles [20].

bond lengths ranging between 1.35(6) to 1.71(5) Å. The μ-H ligand spans the same Os-Os bond edge as the NC_3H_6 ligand but lies on the other side of the Os_3 plane [20].

Thermolysis in refluxing octane for 20 h yielded No. 3 (27%), $(\mu\text{-H})Os_3(\mu_3\text{-}\eta^2\text{-}C_3H_4NC\text{-}cyclo)(CO)_9$ (15%), and $(\mu\text{-H})_3Os_5(\mu_2\text{-}\eta^2\text{-}NC_3H_4)(CO)_{14}$ (see "Organoosmium Compounds" B 8, 1995, Section 5.7; 3%) [20].

References:

[1] Süss-Fink, G. (Z. Naturforsch. **35b** [1980] 454/7).
[2] Adams, R. D.; Katahira, D. A.; Yang, L.-W. (J. Organomet. Chem. **219** [1981] 85/101).
[3] Adams, R. D.; Katahira, D. A.; Yang, L.-W. (J. Organomet. Chem. **219** [1981] 241/9).
[4] Churchill, M. R.; Wasserman, H. J. (Inorg. Chem. **20** [1981] 1580/4).
[5] Dawoodi, Z.; Mays, M. J.; Raithby, P. R. (J. Organomet. Chem. **219** [1981] 103/13).
[6] Dawoodi, Z.; Mays, M. J.; Orpen, A. G. (J. Organomet. Chem. **219** [1981] 251/7).
[7] Laing, M.; Sommerville, P.; Wheatley, P. J.; Mays, M. J.; Dawoodi, Z. (Acta Crystallogr. B **37** [1981] 2230/2).
[8] Banford, J.; Dawoodi, Z.; Henrick, K.; Mays, M. J. (J. Chem. Soc. Chem. Commun. **1982** 554/6).
[9] Burgess, K.; Johnson, B. F. G.; Lewis, J.; Raithby, P. R. (J. Chem. Soc. Dalton Trans. **1982** 2085/92).
[10] Deeming, A. J.; Ghatak, I.; Owen, D. W.; Peters, R. (J. Chem. Soc. Chem. Commun. **1982** 392/3).

[11] Süss-Fink, G.; Khan, L.; Raithby, P. R. (J. Organomet. Chem. **228** [1982] 179/89).
[12] Süss-Fink, G.; Raithby, P. R. (Inorg. Chim. Acta **71** [1983] 109/14).
[13] Dawoodi, Z.; Mays, M. J.; Henrick, K. (J. Chem. Soc. Dalton Trans. **1984** 433/40).

[14] Samkoff, D. E.; Shapley, J. R.; Churchill, M. R.; Wasserman, H. J. (Inorg. Chem. **23** [1984] 397/402).

[15] Banford, J.; Mays, M. J.; Raithby, P. R. (J. Chem. Soc. Dalton Trans. **1985** 1355/60).

[16] Lausarot, P. M.; Vaglio, G. A.; Valle, M.; Tiripicchio, A.; Tiripicchio Camellini, M. (J. Organomet. Chem. **291** [1985] 221/9).

[17] Zuffa, J. L.; Gladfelter, W. L. (J. Am. Chem. Soc. **108** [1986] 4669/71).

[18] Deeming, A. J.; Donovan-Mtunzi, S.; Kabir, S. E.; Arce, A. J.; de Sanctis, Y. (J. Chem. Soc. Dalton Trans. **1987** 1457/61).

[19] Rosenberg, E.; Kabir, S. E.; Hardcastle, K. I.; Day, M.; Wolf, E. (Organometallics **9** [1990] 2214/7).

[20] Adams, R. D.; Chen, G. (Organometallics **12** [1993] 2070/7).

3.1.1.7.2.4 Compounds with μ_3-η^1-Bonded N Ligands

The compounds dealt with in this section have skeletons of the type $Os_3(CO)_n(\mu_3$-η^1-NR) ($n = 8$ to 10) bearing additional hydride and/or halogen ligands. Only Nos. 25 and 26 have two bridging μ_3-η^1-NR moieties. In Table 3, the compounds represented by the general Formulas I to VI are arranged according to the number of the CO groups.

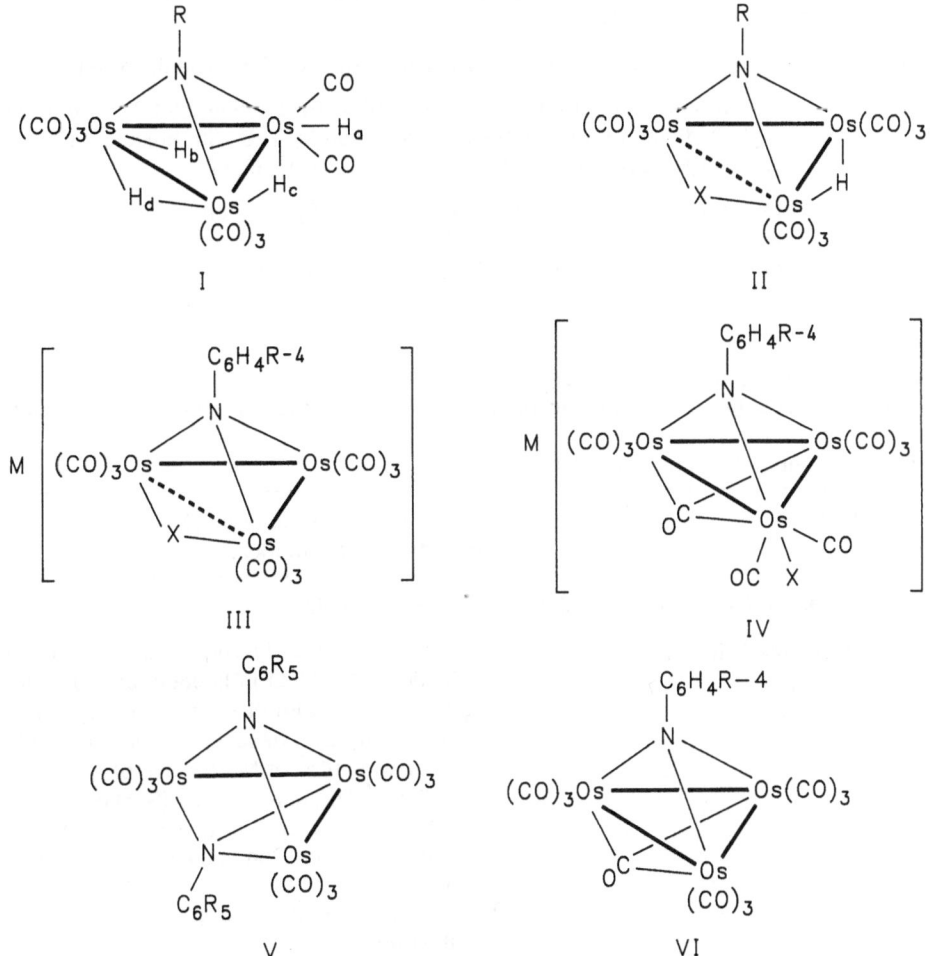

References on p. 58

In general, μ_3-bridging N ligands exhibit a strong shortening influence on Os–Os bonds, which is partially counterbalanced by the bond lengthening effect of bridging hydride ligands. In compounds of type I having three bridging hydrides (see No. 2) the Os–Os bonds are likely equivalent, whereas in compounds containing only two μ-H ligands (see Nos. 4 and 9) the two bridged bond edges are slightly longer than the unbridged Os–Os interaction, which may be described in terms of a two-center two-electron bond [6].

The compounds of the types $(\mu\text{-H})Os_3(CO)_9(\mu_3\text{-}\eta^1\text{-NR})(\mu\text{-X})$ (Formula II) and $M[Os_3(CO)_9\text{-}(\mu_3\text{-}\eta^1\text{-NC}_6H_4R\text{-}4)(\mu\text{-X})]$ (Formula III) have three Os–Os bonds for $X=H$, whereas in the case of $X=Cl$, Br, or I the Os–Os distance bridged by the halogen ligand is elongated, lying in the nonbonding range.

Some compounds listed in Table 3 were prepared by the following methods:

Method I: Compounds of the type $M[Os_3(CO)_9(\mu_3\text{-}\eta^1\text{-NC}_6H_4R\text{-}4)(\mu\text{-X})]$ (Formula III) and
              $M[Os_3(\mu_3\text{-CO})(CO)_8(\mu_3\text{-}\eta^1\text{-NC}_6H_4R\text{-}4)X]$ (Formula IV) were prepared from
              $Os_3(CO)_{10}(\mu_3\text{-}\eta^1\text{-NC}_6H_4R\text{-}4)$ (R=H, CH_3; Nos. 27, 28) and MX in THF for 2.5
              to 3 h. Removal of the solvent yielded either pure $M[Os_3(CO)_9(\mu_3\text{-}\eta^1\text{-}$
              $NC_6H_4R\text{-}4)(\mu\text{-X})]$ for X=Cl, or an isomeric mixture of $M[Os_3(CO)_9(\mu_3\text{-}\eta^1\text{-}$
              $NC_6H_4R\text{-}4)(\mu\text{-X})]$ and $M[Os_3(\mu_3\text{-CO})(CO)_8(\mu_3\text{-}\eta^1\text{-NC}_6H_4R\text{-}4)X]$ for X=Br or I;
              $M=N\{P(C_6H_5)_3\}_2$ [13].

Method II: Compounds of the type $(\mu\text{-H})_2Os_3(CO)_9(\mu_3\text{-}\eta^1\text{-NR})$ (Formula II) were prepared

              a. by thermolysis of $Os_3(CO)_{10}(\mu_2\text{-}\eta^2\text{-NHN=NR})$ (Section 3.1.1.7.3.2) in reflux-
                 ing 1,2-dimethoxyethane for 24 h, followed by TLC with hexane as eluant;
                 $R=n\text{-}C_4H_9$, $CH_2C_6H_5$, cyclo-C_6H_{11}, and C_6H_5 [5, 6], or

              b. by thermolysis of $(\mu\text{-H})Os_3(CO)_9(\mu_2\text{-}\eta^1\text{-NHR})$ (Section 3.1.1.7.2.1) in deca-
                 hydronaphthalene between 187 to 198 °C for 8 to 24 h, or neat at 210 °C
                 in a sealed tube; $R=C_6H_5$, $C_6H_4F\text{-}4$, $C_6H_4CH_3\text{-}4$ [1].

Table 3
Compounds with $\mu_3\text{-}\eta^1$-Bonded N Ligands.
An asterisk preceding the compound number indicates further information at the end of the table, pp. 53/8.
Explanations, abbreviations, and units on p. X.

No. compound	method of preparation (yield) properties and remarks

compounds of the type $(\mu\text{-H})_3HOs_3(CO)_8(\mu_3\text{-}\eta^1\text{-NR})$ (Formula I)

| 1 $(\mu\text{-H})_3HOs_3(CO)_8(\mu_3\text{-}\eta^1\text{-NH})$ | by hydrogenation of $(\mu\text{-H})Os_3(CO)_{10}(\mu_2\text{-}\eta^1\text{-NO})$ (Section 3.1.1.7.2.2) in heptane at 140 °C for 2 h under 136 atm of H_2, followed by chromatography with hexane as eluant (30%), together with 10% of No. 3, 26% of $(\mu\text{-H})Os_3(CO)_{10}(\mu_2\text{-}\eta^1\text{-NH}_2)$ (Section 3.1.1.7.2.1), and 26% of $(\mu\text{-H})_4Os_4(CO)_{12}$ (see "Organoosmium Compounds" B 8, 1995, 4.1.1.3.1) [8] |
| | by hydrogenation of No. 3 as described before (18%) [8] |

Table 3 (continued)

No. compound	method of preparation (yield) properties and remarks
	yellow-orange powder (from hexane) [8] ^{1}H NMR (C_6D_6): -20.40 (t, H_d, $J(H_b,H_d)=2.81$), -18.47 (t, H_c; $J(H_c,H_d)=2.89$), -16.46 (dd, H_b; $J(H_a,H_b)=11.04$, $J(H_b,H_d)=2.84$), -12.74 (dd, H_a; $J(H_a,H_c)=3.18$, $J(H_a,H_b)=11.04$), 4.42 (br s, NH); spectrum depicted [8] IR (hexane): 1962 s, 1992 w, 2021 s, 2032 s, 2040 s, 2060 m, 2080 w, 2101 s, 2120 w (all CO), 3390 vw (NH) [8] FAB mass spectrum: [M]$^+$ [8] carbonylation at 140 °C for 0.5 h under 55 atm of CO gave 40% of No. 3 as the main product [8]
*2 $(\mu-H)_3HOs_3(CO)_8(\mu_3-\eta^1-NCH_2CF_3)$	by hydrogenation of $(\mu-H)Os_3(CO)_{10}(\mu_2-\eta^1-N=CHCF_3)$ (see Section 3.1.1.7.2.3) in hexane at 140 °C for 16 h under 49 atm of H_2 (20%), along with 25% of No. 5 and 30% of $(\mu-H)Os_3(CO)_{10}(\mu_2-\eta^1-NHCH_2CF_3)$ (Section 3.1.1.7.2.1) [4, 7] colorless crystals (from pentane at -5 °C) [4, 7] ^{1}H NMR (CD_2Cl_2): -19.01, -17.85 (s's, OsH$_d$?, OsH$_c$?), -16.11 (d, OsH$_b$; $J(H,H)=10.99$), -12.29 (d, OsH$_a$; $J(H_a,H_b)=9.81$), 4.77 (q, CH_2; $J(H,F)=7.36$) [4, 7] IR (hexane): 1957 m, 1964 w, 2022 m, 2026 sh, 2034 vs, 2043 m, 2045 m, 2104 s, 2122 w [4, 7]; (Nujol mull): 1050 w(br), 1070 w, 1120 s, 1145 sh, 1250 s(br) (all CF) [7] mass spectrum: [M]$^+$ [7]

compounds of the type $(\mu-H)Os_3(CO)_9(\mu_3-\eta^1-NR)(\mu-X)$ (Formula II; X=H, Cl, or I)

3 $(\mu-H)_2Os_3(CO)_9(\mu_3-\eta^1-NH)$	by carbonylation of No. 1 in heptane at 140 °C under 55 atm of CO (40%) [8] by-product in the preparation of No. 1 by hydrogenation of $(\mu-H)Os_3(CO)_{10}(\mu_2-\eta^1-NO)$ (Section 3.1.1.7.2.2) in heptane at 140 °C for 2 h under 136 atm of H_2 (10%) [8]; see also No. 1 pale yellow crystals [8] ^{1}H NMR ($CDCl_3$): -19.44 (s, OsH); (C_6D_6): 4.06 (br s, NH) [8] IR (hexane): 1982 m, 1991 m, 2008 s, 2031 s, 2052 s, 2080 s, 2115 m (all CO), 3383 (NH) [8]

References on p. 58

Table 3 (continued)

No. compound	method of preparation (yield) properties and remarks
3 (continued)	carbonylation in heptane at 140 °C for 30 h under ca. 20 atm of CO yielded $(\mu\text{-H})Os_3\text{-}(CO)_{10}(\mu_2\text{-}\eta^1\text{-}NH_2)$ (Section 3.1.1.7.2.1) [8] hydrogenation in heptane at 140 °C for 2 h under 136 atm of H_2 gave No. 1 and mainly unreacted starting material [8]
*4 $(\mu\text{-H})_2Os_3(CO)_9(\mu_3\text{-}\eta^1\text{-}NCH_3)$	by decarbonylation of $(\mu\text{-H})Os_3(CO)_{10}(\mu_2\text{-}\eta^1\text{-}NHCH_3)$ (Section 3.1.1.7.2.1) or $(\mu\text{-H})Os_3\text{-}(CO)_{10}\{\mu_2\text{-}\eta^2\text{-}N(CH_3)CH{=}O\}$ (Section 3.1.1.7.3.2) at 150 °C [2] ^1H NMR $(CDCl_3)$: -18.74 (OsH), 4.69 (CH_3); the μ-H ligands are equivalent, even at -60 °C [3] IR (hexane): 1974 m, 1986 s, 2005 s, 2032 s, 2078 s, 2082 s, 2112 m (all CO) [3]
*5 $(\mu\text{-H})_2Os_3(CO)_9(\mu_3\text{-}\eta^1\text{-}NCH_2CF_3)$	for formation, see No. 2 [4, 7] by carbonylation of No. 2 in CD_2Cl_2 at 140 °C for 18 h under 50 atm of CO; purified by TLC with CH_2Cl_2/hexane as eluant (25%) [4, 7] pale yellow crystals (from hexane) [7] ^1H NMR (CD_2Cl_2): -18.57 (s, OsH), 4.48 (q, CH_2; J(H,F)=8.56) [7]; see also [4] IR (hexane): 1976 m, 1992 s, 2005 s, 2010 sh, 2034 s, 2039 sh, 2056 vs, 2082 s, 2118 m (all CO) [4, 7]; (Nujol mull): 1145 vs, 1160 s, 1271 m, 1297 w (all CF) [7] mass spectrum: $[M]^+$ [7]
6 $(\mu\text{-H})_2Os_3(CO)_9(\mu_3\text{-}\eta^1\text{-}NC_4H_9\text{-}n)$	IIa [5], IIa (42%) [6] ^1H NMR $(CDCl_3)$: -18.47 (s, OsH), 1.08 to 1.99 (m, CH_3), 4.19 to 4. 23 (m, CH_2) [6] IR (hexane): 1971 w, 1983 m, 2001 s, 2031 m, 2050 s, 2077 s, 2111 w (all CO) [6] EI mass spectrum: $[M]^+$ [6]
7 $(\mu\text{-H})_2Os_3(CO)_9(\mu_3\text{-}\eta^1\text{-}NCH_2C_6H_5)$	IIa [5], IIa (40%) [6] ^1H NMR (CD_2Cl_2): -18.65 (s, OsH), 5.08 (s, CH_2), 7.39 (m, C_6H_5) [6] IR (hexane): 1974 w, 1986 m, 2003 m, 2029 m, 2030 s, 2052 s, 2078 w, 2111 w (all CO) [6] EI mass spectrum: $[M]^+$ [6]
8 $(\mu\text{-H})_2Os_3(CO)_9(\mu_3\text{-}\eta^1\text{-}NC_6H_{11}\text{-}cyclo)$	IIa [5], IIa (38%) [6] ^1H NMR $(CDCl_3)$: -18.53 (s, OsH), 1.09 to 2.32 (m, CH_2), 3.01 (m, CH) [6]

References on p. 58

Table 3 (continued)

No. compound	method of preparation (yield) properties and remarks
	IR (hexane): 1973 w, 1985 m, 2031 s, 2051 s, 2078 s, 2111 w (all CO) [6] EI mass spectrum: [M]$^+$ [6]
*9 (μ-H)$_2$Os$_3$(CO)$_9$(μ_3-η^1-NC$_6$H$_5$)	IIa [5], IIa (81%) [6] IIb (30%) [1]; similar thermolysis in refluxing nonane for 3 h gave initially (μ-H)$_2$Os$_3$-(μ_3-η^2-NHC$_6$H$_4$)(CO)$_9$ (ca. 90%; see "Organoosmium Compounds" B 5, 1994, p. 285), which isomerized partially on further heating for 19 h giving the title compound and the starting material (μ-H)Os$_3$(CO)$_{10}$-(μ_2-η^1-NHC$_6$H$_5$) (Section 3.1.1.7.2.1) [1] orange material, m.p. 174 to 176 °C [1] ^{1}H NMR (CD$_2$Cl$_2$): $-$18.19 (s, OsH), 7.19 (m, C$_6$H$_5$) [6]; (CDCl$_3$): $-$18.33 (s, OsH), 7.12 (m, C$_6$H$_5$) [1] IR (hexane): 1977 m, 1986 s, 2010 s, 2034 s, 2056 s, 2083 s, 2115 m (all CO) [6]; see also [1] EI mass spectrum: [M]$^+$ [6]
10 (μ-H)$_2$Os$_3$(CO)$_9$(μ_3-η^1-NC$_6$H$_4$F-4)	IIb, along with (μ-H)$_2$Os$_3$(μ_3-η^2-NHC$_6$H$_3$F-5)-(CO)$_9$ (see "Organoosmium Compounds" B 5, 1994, p. 286) as a by-product [1] IR (cyclohexane): 1977 m, 1987 s, 2010 s, 2034 s, 2056 s, 2083 s, 2116 m (all CO); slightly contaminated with (μ-H)Os$_3$(CO)$_{10}$-(μ_2-η^1-NHC$_6$H$_4$F-4) (Section 3.1.1.7.2.1) [1]
11 (μ-H)$_2$Os$_3$(CO)$_9$(μ_3-η^1-NC$_6$H$_4$CH$_3$-4)	IIb [1] brown material, m.p. 175.5 to 176.5 °C [1] ^{1}H NMR (CDCl$_3$): $-$18.33 (s, OsH), 2.28 (s, CH$_3$), 6.90, 7.08 (m's of the AA'BB' type, C$_6$H$_4$) [1] IR (cyclohexane): 1976 s, 1986 s, 2009 s, 2032 s, 2036 sh, 2055 s, 2082 s, 2114 s (all CO) [1]
12 (μ-H)Os$_3$(CO)$_9$(μ_3-η^1-NC$_6$H$_5$)(μ-Cl)	by protonation of No. 15 (generated in situ) with CF$_3$CO$_2$H in THF at $-$80 °C under 1 atm of CO, followed by warmup to room temperature and workup by TLC with CH$_2$Cl$_2$/hexane (1:9) as eluant (84%), along with exo/endo-Os$_3$(CO)$_{10}$(μ_2-η^1-NHC$_6$H$_5$)-(μ-Cl) (Section 3.1.1.7.2.1) and (μ-H)Os$_3$(CO)$_{10}$(μ-Cl) (see "Organoosmium Compounds" B 3, 1994, p. 82) [13]

References on p. 58

Table 3 (continued)

No. compound	method of preparation (yield) properties and remarks
12 (continued)	by protonation of [N{P(C_6H_5)_3}_2]-[Os_3{μ_2-η^2-C(O)NC_6H_5}(CO)_{10}(μ-Cl)] (see "Organoosmium Compounds" B 5, 1994, p. 248) under similar conditions (only 3%), together with 5% of endo/exo-Os_3(CO)_{10}-(μ_2-η^1-NHC_6H_5)(μ-Cl) (Section 3.1.1.7.2.1) and 24% of Os_3{μ_2-η^1-C(OH)NC_6H_5}-(CO)_{10}(μ-Cl) (see "Organoosmium Compounds" B 5, 1994, p. 250) [13] pale yellow powder [13] ^{1}H NMR (THF-d_8): −12.19 (s, OsH), 6.98 to 7.52 (m, C_6H_5) [13] ^{13}C NMR (THF-d_8, −73 °C; in the presence of Cr(acac)_3): 127 to 132 (C-2 to 6 of NC_6H_5), 164.4 (s, 1 CO), 167.6 (d, 1 CO of OsH; ^{2}J(C,H) = 17.7), 169.3 (br t, C-1 of NC_6H_5), 172.9 (br, CO), 173.4, 173.8 (s's, each 1 CO), 176.1 (d, 1 CO of OsH; ^{2}J(C,H) = 11.4), 176.4, 182.1, 187.0 (s's, each 1 CO) [13] IR (pentane): 1975 w, 1993 m, 2007 s, 2029 m, 2043 w, 2054 vs, 2083 m, 2117 w (all CO) [13] FAB mass spectrum (18-crown-6/tetraglyme matrix): [M]^+ [13]
13 (μ-H)Os_3(CO)_9(μ_3-η^1-NC_6H_5)(μ-I)	by protonation of No. 23 (generated in situ) with CF_3CO_2H in THF overnight, followed by workup by TLC with CH_2Cl_2/hexane (1:9) as eluant (32%), along with ca. 7% of exo/endo-Os_3(CO)_{10}(μ_2-η^1-NHC_6H_5)(μ-I) (Section 3.1.1.7.2.1); only 22% of No. 13 were obtained upon protonation in the presence of CO [13] by protonation of [N{P(C_6H_5)_3}_2]-[Os_3{μ_2-η^2-C(O)NC_6H_5}(CO)_{10}(μ-I)] (see "Organoosmium Compounds" B 5, 1994, p. 250) at −80 °C under 1 atm of CO (only 5%), together with 5% of endo/exo-Os_3(CO)_{10}(μ_2-η^1-NHC_6H_5)(μ-I) (Section 3.1.1.7.2.1) and 51% of Os_3{μ_2-η^1-C(OH)NC_6H_5}(CO)_{10}(μ-I) (see "Organoosmium Compounds" B 5, 1994, p. 251) [13] pale yellow crystals (from pentane) [13] ^{1}H NMR (CD_2Cl_2): −13.34 (s, OsH), 6.99 to 7.41 (C_6H_5) [13] IR (pentane): 1973 w, 1991 m, 2007 vs, 2027 m, 2040 w, 2052 vs, 2080 s, 2114 w (all CO) [13] EI mass spectrum: [M]^+ [13]

References on p. 58

Table 3 (continued)

No. compound	method of preparation (yield) properties and remarks

compounds of the type M[Os$_3$(CO)$_9$(μ_3-η^1-NC$_6$H$_4$R-4)(μ-X)] (Formula III; X=H, Cl, Br, or I)

14 M[(μ-H)Os$_3$(CO)$_9$(μ_3-η^1-NC$_6$H$_5$)]
    M=[N(C$_2$H$_5$)$_4$]

from [N(C$_2$H$_5$)$_4$][(μ-H)Os$_3$(CO)$_{11}$] and C$_6$H$_5$NO$_2$
in refluxing CH$_3$OH for 3 h; workup by TLC
with pentane/CHCl$_3$ (3:2) as eluant [9]
intermediate in the catalytic formation of
phenyl–isocyanate from nitrobenzene
carbonylation in the presence of polymer-
supported [(μ-H)Os$_3$(μ-CO)(CO)$_{10}$]$^-$ (see
"Organoosmium Compounds" B 3, 1994,
p. 55) [9]
purple solid [9]
^{1}H NMR (CDCl$_3$): $-$18.3 (OsH), 7.26 (C$_6$H$_5$) [9]
IR (CHCl$_3$): 1984 w, 2006 m, 2033 m, 2054 s,
2081 s, 2116 m (all CO) [9]

M=[Y–CH$_2$N(C$_2$H$_5$)$_3$]; Y=poly(styrene–divinylbenzene)

intermediate in the catalytic formation of
phenylisocyanate from nitrobenzene
carbonylation in the presence of
poly(styrene–divinylbenzene)/–
CH$_2$N(C$_2$H$_5$)$_3$–supported [(μ-H)Os$_3$(μ-CO)-
(CO)$_{10}$]$^-$ (not isolated), based on IR
spectroscopy [9]
IR (CHCl$_3$): 1980 w, 2002 m, 2038 m, 2053 s,
2077 s, 2115 m (all CO) [9]

*15 [N{P(C$_6$H$_5$)$_3$}$_2$][Os$_3$(CO)$_9$(μ_3-η^1-NC$_6$H$_5$)(μ-Cl)]

I (68%) [13]
orange microcrystalline powder (from pentane)
[13]
^{13}C NMR (THF-d$_8$, $-$68 °C): 120 to 136 (NC$_6$H$_5$
and PC$_6$H$_5$), 173.4 (t, C–1 of NC$_6$H$_5$;
^{2}J(C,H)=7.9), 176.5, 177.2, 185.2 (s's, each
2 CO), 192.2 (s, 1 CO), 194.4 (s, 2 CO) [13]
IR (THF): 1901 w, 1914 w, 1954 vs, 1988 vs,
2033 vs, 2072 vw (all CO) [13]
negative FAB mass spectrum (18–crown-6/
tetraglyme matrix): [M]$^-$ [13]

16 [N{P(C$_6$H$_5$)$_3$}$_2$][Os$_3$(CO)$_9$(μ_3-η^1-NC$_6$H$_4$CH$_3$-4)(μ-Cl)]

I [13]
^{1}H NMR (CD$_2$Cl$_2$): 2.23 (s, CH$_3$), 6.86 to 7.11
(NC$_6$H$_4$), 7.2 to 7.5 (PC$_6$H$_5$) [13]
IR (THF): 1901 w, 1914 w, 1960 s, 1988 vs,
2033 s, 2050 w (all CO) [13]

References on p. 58

Table 3 (continued)

No. compound	method of preparation (yield) properties and remarks

17　$[N\{P(C_6H_5)_3\}_2][Os_3(CO)_9(\mu_3-\eta^1-NC_6H_5)(\mu-Br)]$

I (74%; as an inseparable 1:1 mixture with No. 21) [13]

yellow–orange microcrystalline powder [13]

IR (THF; in the presence of No. 21): 1900 vw, 1910 vw (both CO of No. 17), 1956 m, 1963 m, 1968 s, 1980 s(sh), 1989 vs, 2010 m, 2033 s, 2042 s, 2072 m (all CO of Nos. 17 and/or 21) [13]

18　$[N\{P(C_6H_5)_3\}_2][Os_3(CO)_9(\mu_3-\eta^1-NC_6H_4CH_3-4)(\mu-Br)]$

I (as an inseparable ca. 1:1 mixture with No. 22) [13]

^1H NMR (CD_2Cl_2): 2.31 (s, CH_3), 6.86 to 7.33 (NC_6H_4) [13]

IR (THF; in the presence of No. 22): 1902 w, 1913 w, 1932 w(sh), 1958 m, 1988 vs, 2010 s, 2033 s, 2041 s, 2072 m, 2084 w(sh) (all CO of No. 18 and/or No. 22) [13]

19　$[N\{P(C_6H_5)_3\}_2][Os_3(CO)_9(\mu_3-\eta^1-NC_6H_5)(\mu-I)]$

I (81%; inseparable mixture with No. 23 as the main product) [13]

yellow–orange microcrystalline powder (from pentane) [13]

for IR and mass spectra, see No. 23 [13]

for protonation, see No. 23 [13]

20　$[N\{P(C_6H_5)_3\}_2][Os_3(CO)_9(\mu_3-\eta^1-NC_6H_4CH_3-4)(\mu-I)]$

I (inseparable mixture with No. 24 as the main product) [13]

^1H NMR (CD_2Cl_2): 2.33 (s, CH_3), 6.84 to 7.11 (NC_6H_4), 7.3 to 7.8 (PC_6H_5) [13]

for IR spectrum, see No. 24 [13]

compounds of the type $M[Os_3(\mu_3-CO)(CO)_8(\mu_3-\eta^1-NC_6H_4R-4)X]$ (Formula IV; X=Cl, Br, or I)

21　$[N\{P(C_6H_5)_3\}_2][Os_3(\mu_3-CO)(CO)_8(\mu_3-\eta^1-NC_6H_5)Br]$

I; see No. 17 [13]

yellow–orange microcrystalline solid (from pentane) [13]

IR (THF; in the presence of No. 17): 1665 vw (μ_3-CO); for the other CO resonances, see No. 17 [13]

22　$[N\{P(C_6H_5)_3\}_2][Os_3(\mu_3-CO)(CO)_8(\mu_3-\eta^1-NC_6H_4CH_3-4)Br]$

I; see No. 18 [13]

yellow–orange microcrystals (from pentane) [13]

References on p. 58

Table 3 (continued)

No. compound	method of preparation (yield) properties and remarks

¹H NMR (CD₂Cl₂): 2.22 (s, CH₃), 6.88 to 7.14

¹H NMR (CD$_2$Cl$_2$): 2.22 (s, CH$_3$), 6.88 to 7.14 (NC$_6$H$_4$) [13]

IR (THF; in the presence of No. 18): 1671 vw (μ$_3$-CO); for the other CO resonances, see No. 18 [13]

23 [N{P(C$_6$H$_5$)$_3$}$_2$][Os$_3$(μ$_3$-CO)(CO)$_8$(μ$_3$-η^1-NC$_6$H$_5$)I]

I; see No. 19 [13]

yellow-orange microcrystalline powder (from pentane) [13]

¹³C NMR (THF-d$_8$, −73 °C; in the presence of No. 19): 122 to 136 (C-2 to 6 of NC$_6$H$_5$ and PC$_6$H$_5$), 165.0 (br, C-1 of NC$_6$H$_5$), 173.9, 174.1, 176.5, 176.8, 179.8, 181.1, 181.8, 184.0, (s's, each 1 CO), 234.4 (s, μ$_3$-CO) [13]

IR (THF): 1657 vw (μ$_3$-CO), 1935 m, 1970 s(sh), 1979 vs, 2008 vs, 2041 vs, 2070 m (all CO) [13]

negative FAB mass spectrum (18-crown-6/ tetraglyme matrix): [M]$^-$ [13]

protonation (in the presence of minor amounts of No. 19) with CF$_3$CO$_2$H in THF overnight gave 32% No. 13 and ca. 7% of endo/exo-Os$_3$(CO)$_{10}$(μ$_2$-η^1-NHC$_6$H$_5$)(μ-I) (Section 3.1.1.7.2.1); only 22% of No. 13 was obtained upon protonation in the presence of CO [13]

carbonylation of No. 23 (generated in situ) in THF under 1 atm of CO led to a mixture of [N{P(C$_6$H$_5$)$_3$}$_2$][Os$_3${μ$_2$-η^2-C(O)NC$_6$H$_5$}-(CO)$_{10}$(μ-I)] (see "Organoosmium Compounds" B 5, 1994, p. 250) and [N{P(C$_6$H$_5$)$_3$}$_2$][Os$_3$(CO)$_{11}$I] (see "Organo-osmium Compounds" B 3, 1994, p. 76) [13]

24 [N{P(C$_6$H$_5$)$_3$}$_2$][Os$_3$(μ$_3$-CO)(CO)$_8$(μ$_3$-η^1-NC$_6$H$_4$CH$_3$-4)I]

I; see No. 20; [N{P(C$_6$H$_5$)$_3$}$_2$]-[Os$_3${μ$_2$-η^2-C(O)NC$_6$H$_4$CH$_3$-4}(CO)$_9$(μ-I)] (see "Organoosmium Compounds" B 5, 1994, p. 250) was formed as a co-product [13]

¹H NMR (CD$_2$Cl$_2$): 2.25 (s, CH$_3$), 6.86 to 7.06 (NC$_6$H$_4$), 7.3 to 7.8 (PC$_6$H$_5$) [13]

IR (THF): 1659 w(br) (μ$_3$-CO of No. 24), 1901 vw, 1914 vw, 1935 w(sh), 1971 w(sh), 1984 vs, 1993 vs, 2007 s, 2032 m(sh), 2041, 2049 m(sh), 2070 m, 2083 vw (all CO of No. 24 and/or No. 20) [13]

References on p. 58

Table 3 (continued)

No. compound	method of preparation (yield) properties and remarks

compounds of the type Os$_3$(CO)$_9$(μ_3-η^1-NC$_6$R$_5$)$_2$ (Formula V)

25 Os$_3$(CO)$_9$(μ_3-η^1-NC$_6$H$_5$)$_2$

from Os$_3$(CO)$_{12}$ and an excess of C$_6$H$_5$NO in octane at 126 °C for 24 h under 3.5 atm of CO in a sealed tube, followed by column chromatography with hexane as eluant (34%) [10]

from Os$_3$(CO)$_{12}$ and C$_6$H$_5$NO (1:1) in octane at 126 °C for 7 h, followed by chromatography with CH$_2$Cl$_2$/hexane (1:4) as eluant (19%), along with 5% of No. 27 [10]

orange crystals [10]

^{1}H NMR (CD$_2$Cl$_2$): 6.76, 6.79, 6.90, 6.92, 6.94, 7.09, 7.12, 7.14 (all C$_6$H$_5$) [10]

IR (hexane): 1972 w, 1980 m, 2001 m, 2011 s, 2051 s, 2074 s, 2095 w (all CO) [10]

*26 Os$_3$(CO)$_9$(μ_3-η^1-NC$_6$F$_5$)$_2$

from Os$_3$(CO)$_{12}$ and C$_6$F$_5$NO (1:1) in refluxing octane for 3 h; purified by TLC with CH$_2$Cl$_2$/hexane (1:9) as eluant (no yield given) [12]

red crystals (from hexane/CH$_2$Cl$_2$) [12]

^{19}F NMR (CDCl$_3$; vs. internal CF$_3$CO$_2$H): −85.95 (m, 4 F), −84.8 (m, 3 F), −72.11 (m, 1 F), −65.90 (m, 2 F) [12]

IR (cyclohexane): 998 (NC), 1491, 1512 (C$_6$F$_5$), 1992 w, 2005 m, 2014 vs, 2018 s(sh), 2035 m, 2062 vs, 2068 (sh) m, 2083 vs, 2104 w (all CO) [12]

the HPLC separation from Os$_3$(CO)$_{11}$\{μ_2-η^2-N(C$_6$F$_5$)O\} (Section 3.1.1.7.3.2), Os$_3$(CO)$_{11}$NCCH$_3$, and Os$_3$(CO)$_{12}$ was investigated [14]

compounds of the type Os$_3$(μ_3-CO)(CO)$_9$(μ_3-η^1-NC$_6$H$_4$R-4) (Formula VI)

27 Os$_3$(μ_3-CO)(CO)$_9$(μ_3-η^1-NC$_6$H$_5$)

from Os$_3$(CO)$_{11}$NCCH$_3$ and C$_6$H$_5$NO (ca. 1:1) in THF at room temperature for 30 min, followed by column chromatography with CH$_2$Cl$_2$/hexane (1:9) as eluant (54%), along with Os$_3$(CO)$_{12}$ and other unidentified by-products [13]

from Os$_3$(CO)$_{11}$NCCH$_3$ generated in situ and C$_6$H$_5$NO in THF as before, followed by chromatography with hexane as eluant (40%); attempts to observe the presumed nitroso-coordinated intermediates at low temperature failed [11]

References on p. 58

Table 3 (continued)

No. compound	method of preparation (yield) properties and remarks
	for formation, see also No. 25 [10] yellow microcrystals [10, 13] ^{13}C NMR (THF-d$_8$, -63 °C): 123 to 131 (C-2 to 6 of NC$_6$H$_5$), 165.2 (t, C-1 of NC$_6$H$_5$; ^{2}J(C,H) = 7.0), 172.3 (s, 6 CO), 176.2 (s, 3 CO), 247.0 (s, μ_3-CO) [13] IR (hexane): 1695 m, 1995 m, 2002 m, 2020 sh, 2028 s, 2074 s, 2105 w (all CO) [10]; see also [11] mass spectrum: [M]$^+$ [10] reaction with [N{P(C$_6$H$_5$)$_3$}$_2$]X (X=Cl, Br, I) in THF yielded compounds of type III and IV; see Preparation Method I [13]
28 Os$_3$(μ_3-CO)(CO)$_9$(μ_3-η^1-NC$_6$H$_4$CH$_3$-4)	from Os$_3$(CO)$_{11}$NCCH$_3$ and 4-CH$_3$C$_6$H$_4$NO as for No. 27 (51%) [13] ^{1}H NMR (CD$_2$Cl$_2$): 2.30 (s, CH$_3$), 6.9 to 7.2 (NC$_6$H$_4$) [13] ^{13}C NMR (THF-d$_8$, 23 °C): 170.8 (s, 6 equatorial CO), 174.2 (s, 3 axial CO), 244 (s, μ_3-CO) [13] IR (C$_5$H$_{12}$): 1692 w, 2002 w, 2024 vs, 2072 vs, 2103 w (all CO) [13] reaction with [N{P(C$_6$H$_5$)$_3$}$_2$]X (X=Cl, Br, I) in THF yielded compounds of type III and IV; see Preparation Method I [13]

*Further information:

(μ-H)$_3$HOs$_3$(CO)$_8$(μ_3-η^1-NCH$_2$CF$_3$) (Table 3, No. 2) crystallizes in the monoclinic space group C2/c – C$_{2h}^6$ (No. 15) with a = 17.387(3), b = 8.137(2), c = 26.142(3) Å, β = 99.15(4)°; Z = 8, D$_c$ = 3.258 g/cm^3. The structure, shown in **Fig. 18**, is closely related to that of No. 5. The three bridging hydride ligands were not located crystallographically, but believed to span the three Os-Os bond edges as concluded from the three equal metal-metal distances. The terminal hydride ligand, located at the Os(2) center, is disordered between the sites occupied by C′ and C″, resulting in a half occupancy of these sites by CO groups. The disorder is accompanied by a slight disorder in the CF$_3$ group. Average Os-C and C-O distances are 1.892 and 1.172 Å, respectively [4, 7].

Carbonylation under 50 atm of CO in CD$_2$Cl$_2$ at 140 °C gave No. 5 [4, 7].

(μ-H)$_2$Os$_3$(CO)$_9$(μ_3-η^1-NCH$_3$) (Table 3, No. 4) crystallizes in the orthorhombic space group Pmcn (Pnma) – D$_{2h}^{16}$ (No. 62) with a = 14.113(2), b = 6.605(1), c = 17.683(4) Å; Z = 4, D$_c$ = 3.44 g/cm^3. The structure is shown in **Fig. 19**. One of the Os$_3$ faces of the electronically saturated 48-electron cluster is triply bridged by the NCH$_3$ ligand; the N-Os$_3$ mean-plane distance amounts to 1.28(1) Å. The hydride ligands, observed crystallographically, are positioned on the opposite side of the Os$_3$ plane than the NCH$_3$ unit. They bridge the Os(1)-Os(3) bond edges asymmetrically with shorter bond lengths to Os(3) than to Os(1) [3].

References on p. 58

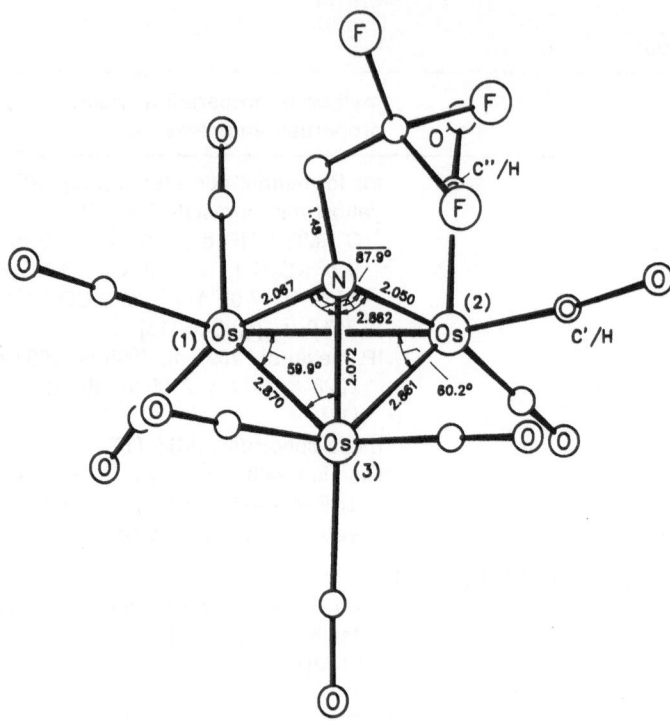

Fig. 18. Molecular structure of $(\mu-H)_3HOs_3(CO)_8(\mu_3-\eta^1-NCH_2CF_3)$ (No. 2) with selected bond distances (in Å) and bond angles [4, 7].

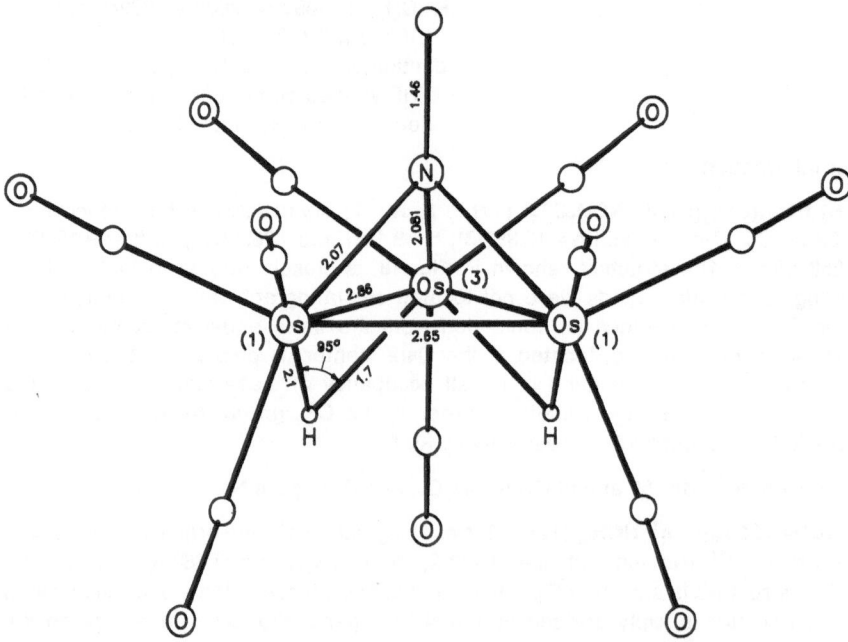

Fig. 19. Molecular structure of $(\mu-H)_2Os_3(CO)_9(\mu_3-\eta^1-NCH_3)$ (No. 4) with selected bond distances (in Å) and a bond angle [3].

References on p. 58

Thermolysis in decahydronaphthalene at 198 °C gave $Os_4(CO)_{12}(\mu_3-\eta^1-NCH_3)$ (see "Organoosmium Compounds" B 8, 1995, Section 4.1.1.3.2) in 35% yield [3, 8].

$(\mu-H)_2Os_3(CO)_9(\mu_3-\eta^1-NCH_2CF_3)$ (Table 3, No. 5) crystallizes in the monoclinic space group $P2_1/c-C_{2h}^5$ (No. 14) with a = 9.383(2), b = 12.388(2), c = 16.171(3) Å, β = 97.79(2)°; Z = 4, D_c = 3.279 g/cm³. The structure, shown in **Fig. 20** [4, 7], is closely related to that of No. 2 and of No. 4. The μ-H-bridged Os(1)-Os(3) and Os(2)-Os(3) distances are longer by ca. 0.14 Å than the remaining Os(1)-Os(2) bond. Both the hydrides and the methylene hydrogens were located crystallographically [7]. Average Os-C and C-O distances are 1.915 and 1.139 Å, respectively [4, 7]. Probably, the bond distances for the Os(1)-Os(2) and Os(1)-Os(3) edges have erroneously been changed in [4] and [7]; the corrected values are given in Fig. 20.

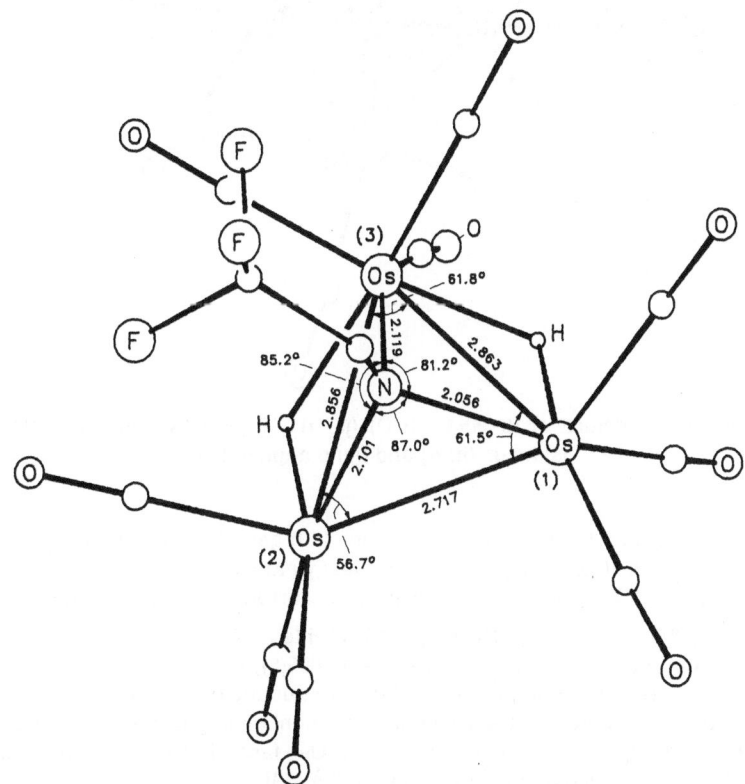

Fig. 20. Molecular structure of $(\mu-H)_2Os_3(CO)_9(\mu_3-\eta^1-NCH_2CF_3)$ (No. 5) with selected bond distances (in Å) and bond angles [4, 7].

$(\mu-H)_2Os_3(CO)_9(\mu_3-\eta^1-NC_6H_5)$ (Table 3, No. 9) crystallizes in the monoclinic space group $P2_1/c-C_{2h}^5$ (No. 14) with a = 11.630(4), b = 9.217(4), c = 18.527(7) Å, β = 99.71(2)°; Z = 4, D_c = 3.11 g/cm³ [5, 6]. The structure of the 48-electron cluster is shown in **Fig. 21**. It is quite similar to that of No. 4. The NC_6H_5 ligand symmetrically caps one face of the Os_3 framework. The phenyl ring forms an angle of 93.5° with the Os_3 triangle. Both hydrides were not located directly but the distribution of the carbonyl ligands indicates that they bridge the two elongated Os(1)-Os(2) and Os(2)-Os(3) bond edges, and that they lie on the opposite side of the Os_3 plane than the capping NC_6H_5 unit [5, 6].

References on p. 58

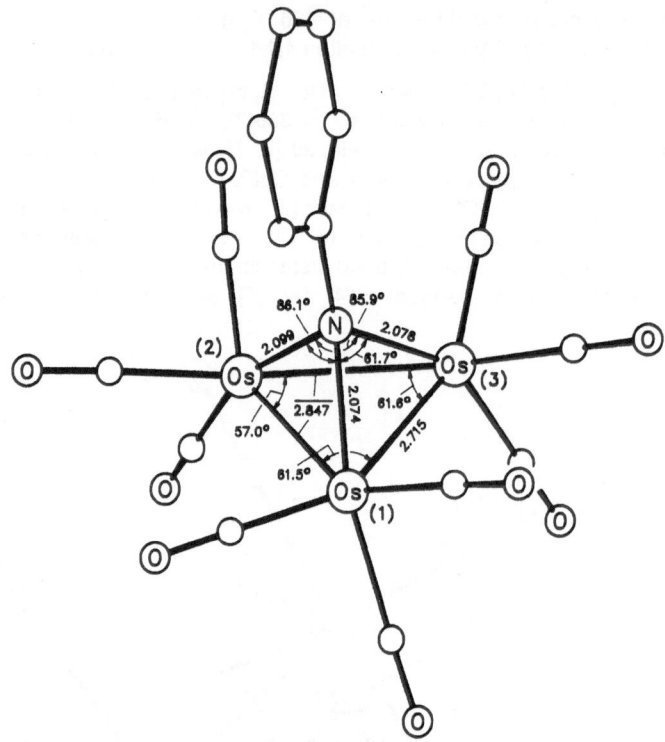

Fig. 21. Molecular structure of $(\mu-H)_2Os_3(CO)_9(\mu_3-\eta^1-NC_6H_5)$ (No. 9) with selected bond distances (in Å) and bond angles [5, 6].

Carbonylation under 1 atm of CO at 175 °C for 14 h gave $(\mu-H)Os_3(CO)_9(\mu_2-\eta^1-NHC_6H_5)$ (Section 3.1.1.7.2.1) in 10 to 20% yield. The product yield increased to ca. 50% upon similar carbonylation at 195 °C for 10 h; no carbonylation occurred at only 150 °C [1].

$[N\{P(C_6H_5)_3\}_2][Os_3(CO)_9(\mu_3-\eta^1-NC_6H_5)(\mu-Cl)]$ (Table **3**, No. **15**) crystallizes in the monoclinic space group $P2_1/c-C_{2h}^5$ (No. 14) with a = 19.685(3), b = 15.258(2), c = 18.650(3) Å, β = 117.744(13)°; Z = 4, D_c = 1.993 g/cm³. The molecular structure of the anion is shown in **Fig. 22**. The cluster possesses only two Os-Os bonds; the nonbonding Os(1)···Os(2) edge is symmetrically bridged by the μ-Cl ligand. One of the Os_3 faces is capped by the phenylimido ligand; the three Os-N bond lengths average 2.108 Å [13].

Carbonylation with one atmosphere of CO in THF at room temperature for 93 h in a sealed tube yielded an inseparable mixture of $[N\{P(C_6H_5)_3\}_2][Os_3\{\mu_2-\eta^2-C(O)NC_6H_5\}-(CO)_{10}(\mu-Cl)]$ (see "Organoosmium Compounds" B 5, 1994, p. 248) and $[N\{P(C_6H_5)_3\}_2]-[Os_3(CO)_{11}Cl]$ ("Organoosmium Compounds" B 3, 1994, p. 75) [13].

Protonation with CF_3CO_2H or $HBF_4 \cdot O(C_2H_5)_2$ under one atmosphere of CO gave No. 12 as the major product, besides endo/exo-$Os_3(CO)_{10}(\mu_2-\eta^1-NHC_6H_5)(\mu-Cl)$ (Section 3.1.1.7.2.1) and $(\mu-H)Os_3(CO)_{10}(\mu-Cl)$ (see "Organoosmium Compounds" B 3, 1994, p. 82) [13].

$Os_3(CO)_9(\mu_3-\eta^1-NC_6F_5)_2$ (Table **3**, No. **26**) crystallizes in the triclinic space group $P\bar{1}-C_i^1$ (No. 2) with a = 9.062(3), b = 12.431(4), c = 13.446(6) Å, α = 90.8°, β = 109.21°, γ = 109.65°; Z = 4,

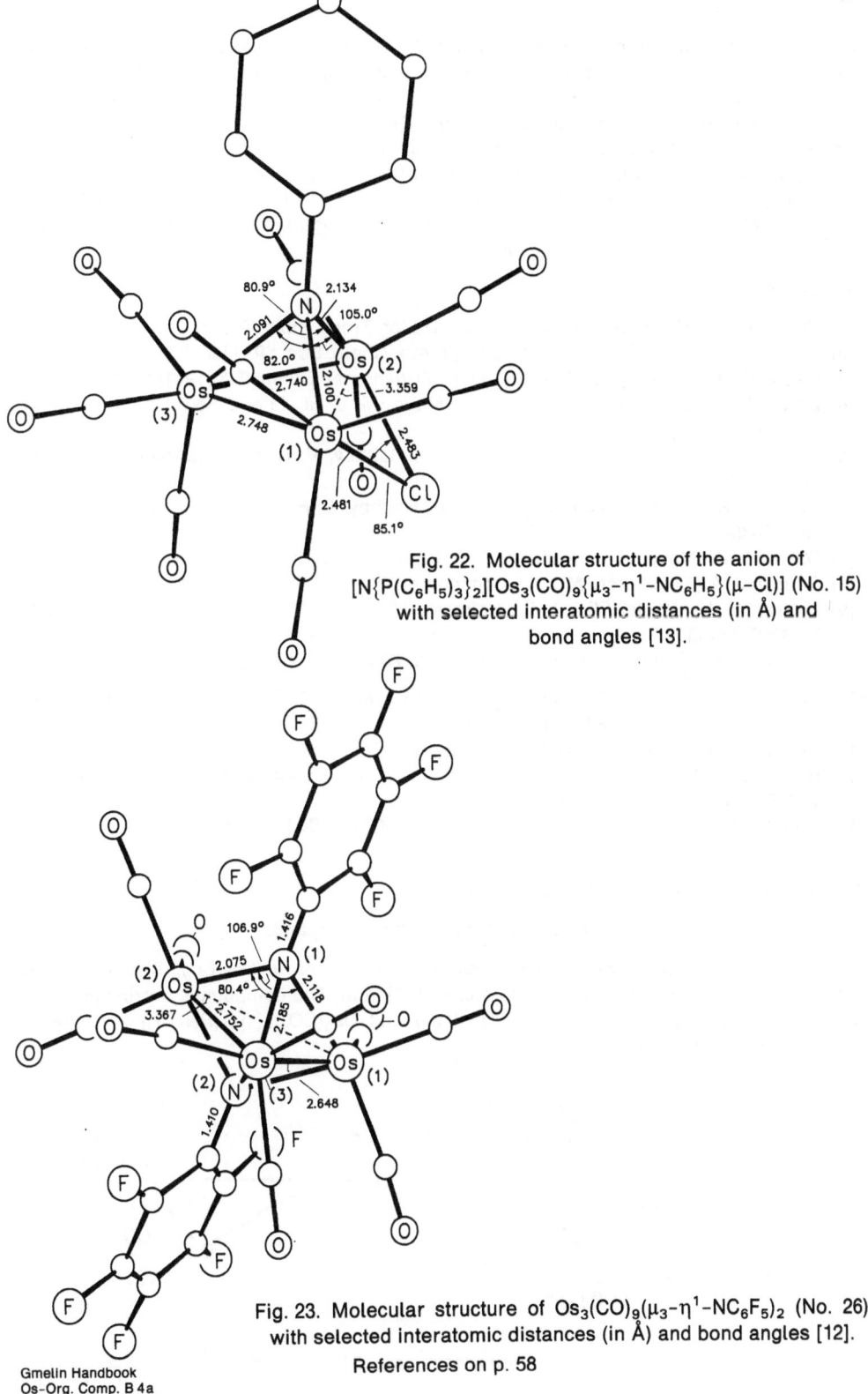

Fig. 22. Molecular structure of the anion of
[N{P(C₆H₅)₃}₂][Os₃(CO)₉{μ₃-η¹-NC₆H₅}(μ-Cl)] (No. 15)
with selected interatomic distances (in Å) and
bond angles [13].

Fig. 23. Molecular structure of Os₃(CO)₉(μ₃-η¹-NC₆F₅)₂ (No. 26)
with selected interatomic distances (in Å) and bond angles [12].

References on p. 58

$D_c = 2.950$ g/cm^3. The structure is shown in **Fig. 23**. The Os$_3$ triangle is capped by two penta-fluorophenylimido ligands above and below the plane. The Os(1)-Os(2) distance of 3.367 Å is in the nonbonding range. The Os-N(1) and Os-N(2) bonds show little variation except for Os(3)-N(1), which is slightly longer than the Os(3)-N(2) bond of 2.157 Å. The C$_6$F$_5$ rings attached to N(1) and N(2) are inclined by 88.5° and 95.9°, respectively, with respect to the Os$_3$ plane. Furthermore, the dihedral angle between the two mean planes containing C$_6$F$_5$ groups is 7.9°, indicating that the two C$_6$F$_5$ rings are not symmetrical with respect to the Os$_3$ plane. The Os-CO and C-O bond lengths range between 1.89 and 1.94 Å, and between 1.109 and 1.140 Å, respectively [12].

References:

[1] Yin, C. C.; Deeming, A. J. (J. Chem. Soc. Dalton Trans. **1974** 1013/7).
[2] Lin, Y. C.; Knobler, C. B.; Kaesz, H. D. (J. Am. Chem. Soc. **103** [1981] 1216/8).
[3] Lin, Y. C.; Knobler, C. B.; Kaesz, H. D. (J. Organomet. Chem. **213** [1981] C 41/C 43).
[4] Banford, J.; Dawoodi, Z.; Henrick, K.; Mays, M. J. (J. Chem. Soc. Chem. Commun. **1982** 554/6).
[5] Burgess, K.; Johnson, B. F. G.; Lewis, J.; Raithby, P. R. (J. Organomet. Chem. **224** [1982] C 40/C 44).
[6] Burgess, K.; Johnson, B. F. G.; Lewis, J.; Raithby, P. R. (J. Chem. Soc. Dalton Trans. **1982** 2085/92).
[7] Dawoodi, Z.; Mays, M. J.; Henrick, K. (J. Chem. Soc. Dalton Trans. **1984** 433/40).
[8] Smieja, J. A.; Gladfelter, W. L. (J. Organomet. Chem. **297** [1985] 349/59).
[9] Marrakchi, H.; Nguini Effa, J.-B.; Haimeur, M.; Lieto, J.; Aune, J. P. (Bull. Soc. Chim. Fr. **1985** 390/2).
[10] Smieja, J. A.; Gladfelter, W. L. (Inorg. Chem. **25** [1986] 2667/70).

[11] Han, S.-H.; Song, J.-S.; Macklin, P. D.; Nguyen, S. T.; Geoffroy, G. L.; Rheingold, A. L. (Organometallics **8** [1989] 2127/38).
[12] Ang, H. G.; Kwik, W. L.; Ong, K. K. (J. Fluorine Chem. **56** [1992] 45/9).
[13] Ramage, D. L.; Geoffroy, G. L.; Rheingold, A. L.; Haggerty, B. S. (Organometallics **11** [1992] 1242/55).
[14] Ang, H. G.; Kwik, W. L.; Ong, K. K. (J. Fluorine Chem. **60** [1993] 43/8).

3.1.1.7.3 Compounds with η²-Bonded N Ligands

3.1.1.7.3.1 Compounds with η²-Bonded Nonbridging N Ligands

The compounds dealt with in this section all consist of an Os$_3$(CO)$_n$ skeleton having a η²-nonbridging four-electron donating chelate ligand. The compounds corresponding to the general types Os$_3$(CO)$_{10}$(η²-RN=CHC$_5$H$_4$N) and Os$_3$(CO)$_{10}${η²-(RN=CH)$_2$} are represented by Formulas I and II.

I

II

References on p. 64

The structure for the compounds of type I and type II appeared to be quite similar, exhibiting the pyridine and imine nitrogens likewise in an axial and equatorial coordination site, as established for $Os_3(CO)_{10}\{\eta^2\text{-}(i\text{-}C_3H_7N=CH)_2\}$ (No. 5). Based on the general observations that sterically bulky alkyl groups prefer equatorial coordination sites, it is assumed that the imine nitrogens bearing the bulky R moieties reside in the equatorial positions, while the pyridine nitrogens are axial. For the CO groups of the compounds of type I a fluxional behavior similar to that observed for the compounds of type II was expected, but variable-temperature ^{13}C NMR determinations failed because of the low solubility of these compounds [1].

The compounds listed in Table 4 were prepared by the following methods:

Method I: Compounds of the type $Os_3(CO)_{10}(\eta^2\text{-}RN=CHC_5H_4N)$ (Formula I) were prepared from $Os_3(CO)_{10}(NCCH_3)_2$ (see "Organoosmium Compounds" B 3, 1994, p. 218) and $RN=CHC_5H_4N$ (ca. 1:1) in $C_6H_5CH_3$ at $-20\,°C$, followed by warmup within 2 h and further stirring for 4 h at room temperature. The products were purified by column chromatography with ether/hexane (2:1) as eluant. Lower yields were obtained upon starting at room temperature [1]; see also [2].

Method II: Compounds of the type $Os_3(CO)_{10}\{\eta^2\text{-}(RN=CH)_2\}$ (Formula II) were prepared from $Os_3(CO)_{10}(NCCH_3)_2$ (In situ generated; see "Organoosmium Compounds" B 3, 1994, p. 218) and an excess of $(RN=CH)_2$ in CH_3CN at $45\,°C$, or alternatively with 1 equivalent of $Os_3(CO)_{10}(NCCH_3)_2$ in THF for 6 h, followed by column chromatography with hexane/ether (10:1) as eluant; $R = i\text{-}C_3H_7$ or $neo\text{-}C_5H_{11}$. Preparations in $C_6H_5CH_3$ led to a considerable decrease in the yield of $Os_3(CO)_{10}\{\eta^2\text{-}(i\text{-}C_3H_7N=CH)_2\}$ (No. 5), while the yield of $Os_3(CO)_{10}\{\eta^2\text{-}(neo\text{-}C_5H_{11}N=CH)_2\}$ (No. 6) is independent of the solvent [1, 3].

Treatment of $Os_3(CO)_{10}(NCCH_3)_2$ with $(cyclo\text{-}C_3H_5N=CH)_2$ under the conditions of Preparation Method II gave yellow $Os_3\{\mu_2\text{-}\eta^3\text{-}CH(NC_3H_5\text{-}cyclo)CH=NC_3H_5\text{-}cyclo\}(CO)_{10}$ (see "Organoosmium Compounds" B 5, 1994, p. 269) and only trace amounts of a red compound, probably $\mathbf{Os_3(CO)_{10}\{\eta^2\text{-}(cyclo\text{-}C_3H_5N=CH)_2\}}$, which converted into the above mentioned yellow complex [3].

The formation mechanisms for $Os_3(CO)_{10}(\eta^2\text{-}RN=CHC_5H_4N)$ (Nos. 1 to 4) and $Os_3(CO)_{10}\{\eta^2\text{-}(RN=CH)_2\}$ (Nos. 5 and 6) are quite similar and discussed in detail [1].

References on p. 64

Table 4

Compounds with η^2-Bonded Nonbridging N Ligands.

An asterisk preceding the compound number indicates further information at the end of the table, pp. 62/4.

Explanations, abbreviations, and units on p. X.

No. compound	method of preparation (yield) properties and remarks

compounds of the type $Os_3(CO)_{10}(\eta^2$-RN=CHC$_5$H$_4$N) (Formula I)

1   $Os_3(CO)_{10}(\eta^2$-i-C$_3$H$_7$N=CHC$_5$H$_4$N)

I (45%), along with ca. 2% of $Os_3\{\mu_2$-η^3-CHN(C$_3$H$_7$-i)C$_5$H$_4$N}(CO)$_{10}$ (see "Organo-osmium Compounds" B 5, 1994, p. 270) [1]; see also [2]

crystals (from ether/hexane at -80 °C) [1]

^{1}H NMR (C$_6$D$_6$): 0.92, 1.06 (d's, each 1 CH$_3$), 4.46 (sept, CH of i-C$_3$H$_7$), 6.01 (dd, H-5), 6.62 (dd, H-4), 6.77 (d, H-3), 7.94 (s, N=CH), 9.29 (d, H-6); $^3J=6$ or 7 [1]

^{13}C NMR (CD$_2$Cl$_2$): 21.4, 29.9, 67.8 (all i-C$_3$H$_7$), 124.3 (C-5), 128.1 (C-4), 135.6 (C-3), 155.1 (C-6), 156.2 (N=CH), 156.3 (C-2); no resonances for CO given [1]

IR (hexane): 1922, 1960, 1978, 1984, 1997, 2014, 2038, 2086 (all CO) [1]

FD mass spectrum (no medium given): [M]$^+$ [1]

thermolysis in refluxing hexane for 24 h gave $(\mu$-H)Os$_3(\mu_2$-η^3-C$_5$H$_3$NCH=NC$_3$H$_7$-i)(CO)$_9$ (see "Organoosmium Compounds" B 5, 1994, p. 265) [1]; see also [2]

2   $Os_3(CO)_{10}(\eta^2$-cyclo-C$_3$H$_5$N=CHC$_5$H$_4$N)

I (18%), along with ca. 2% of $Os_3\{\mu_2$-η^3-CHN(C$_3$H$_5$-cyclo)C$_5$H$_4$N}(CO)$_{10}$ (see "Organoosmium Compounds" B 5, 1994, p. 270) [1]; see also [2]

crystals (from ether/hexane at -80 °C) [1]

^{1}H NMR (CD$_2$Cl$_2$): 1.27 (m, CH$_2$ of C$_3$H$_5$), 4.06 (m, CH of C$_3$H$_5$), 7.17 (dd, H-5), 7.82 (dd, H-4), 8.01 (d, H-3), 8.79 (s, N=CH), 9.40 (d, H-6); $^3J=7$ or 8 [1]

^{13}C NMR (CD$_2$Cl$_2$): 11.2, 13.0, 50.1 (all C$_3$H$_5$), 124.2 (C-5), 128.1 (C-4), 135.4 (C-3), 155.6 (C-2,6), 156.2 (N=CH) [1]

IR (hexane): 1922, 1960, 1978, 1984, 1997, 2014, 2038, 2086 (all CO) [1]

FD mass spectrum (no medium given): [M]$^+$ [1]

FAB mass spectrum (3-mercapto-1,2-propanediol): [M]$^+$, [M$^+$ $-$ n CO]$^+$, n = 1 to 8 [1]

References on p. 64

Table 4 (continued)

No. compound	method of preparation (yield) properties and remarks

3 $Os_3(CO)_{10}(\eta^2\text{-}t\text{-}C_4H_9N\text{=}CHC_5H_4N)$

I (36%) [1]; see also [2]

crystals (from ether/hexane at $-80\,°C$) [1]

^1H NMR (CD_2Cl_2): 1.63 (s, $t\text{-}C_4H_9$), 7.18 (dd, H-5), 7.90 (dd, H-4), 8.07 (d, H-3), 8.97 (s, N=CH), 9.42 (d, H-6); $^3J=6$ or 7 [1]

^{13}C NMR (CD_2Cl_2): 32.7, 68.7 (both $t\text{-}C_4H_9$), 124.6 (C-5), 128.8 (C-4), 136.5 (C-3), 156.3 (C-6), 157.2 (C-2), 159.3 (N=CH); no resonances for CO given [1]

IR (hexane): 1917, 1960, 1971, 1983, 1997, 2011, 2037, 2084 (all CO) [1]

FD mass spectrum (no medium given): $[M]^+$ [1]

thermolysis in refluxing hexane for 24 h gave $(\mu\text{-}H)Os_3(\mu_2\text{-}\eta^3\text{-}C_5H_3NCH\text{=}NC_4H_9\text{-}t)(CO)_9$ ("Organoosmium Compounds" B 5, 1994, p. 266) [1]; see also [2]

4 $Os_3(CO)_{10}(\eta^2\text{-}neo\text{-}C_5H_{11}N\text{=}CHC_5H_4N)$

I (45%) [1]; see also [2]

crystals (from ether/hexane at $-80\,°C$) [1]

^1H NMR (CD_2Cl_2): 1.09 (s, CH_3), 3.97, 4.36 (d's, each 1 H of CH_2; $^2J=12$), 7.17 (dd, H-5; $^3J=6$), 7.81 (dd, H-4; $^3J=7$), 8.02 (d, H-3; $^3J=7$), 8.57 (s, N=CH), 9.43 (d, H-6; $^3J=6$) [1]

^{13}C NMR (CD_2Cl_2): 28.1, 34.2, 78.8 (all $neo\text{-}C_5H_{11}$), 123.9 (C-5), 127.4 (C-4), 135.1 (C-3), 155.2 (C-2), 156.0 (C-6), 159.9 (N=CH); no resonances for CO given [1]

IR (hexane): 1918, 1962, 1971, 1983, 1997, 2012, 2036, 2085 (all CO) [1]

FD mass spectrum (no medium given): $[M-CO]^+$ [1]

compounds of the type $Os_3(CO)_{10}\{\eta^2\text{-}(RN\text{=}CH)_2\}$ (Formula II)

*5 $Os_3(CO)_{10}\{\eta^2\text{-}(i\text{-}C_3H_7N\text{=}CH)_2\}$

II (30%), along with small amounts of $Os_3\{\mu_2\text{-}\eta^3\text{-}CH(NC_3H_7\text{-}i)CH\text{=}NC_3H_7\text{-}i\}(CO)_{10}$ (see "Organoosmium Compounds" B 5, 1994, p. 269) [1, 3]

red crystals (from hexane/ether at $-80\,°C$); stable in solution at room temperature [3]

^1H NMR (C_6D_6): 1.02, 1.08 (d's, each 2 CH_3), 4.58 (sept, CH of $i\text{-}C_3H_7$), 7.74 (s, N=CH); $(CD_2Cl_2, -80\,°C)$: 1.30, 1.32 (d's, each 2 CH_3), 4.54 (sept, CH of $i\text{-}C_3H_7$), 8.41 (s, N=CH); all $^3J=6$; the CH_3 groups are diastereotopic [3]

References on p. 64

Table 4 (continued)

No. compound	method of preparation (yield) properties and remarks
*5 (continued)	^{13}C NMR (CD_2Cl_2, -80 °C): 21.3, 27.0 (each 2 CH_3), 65.4 (CH of i-C_3H_7), 144.4 (N=CH), 169.8 ($CO_{d,d'}$), 174.3 ($CO_{e,e'}$), 182.9 ($CO_{c,c'}$), 184.0 ($CO_{b,b'}$), 188.2 ($CO_{a,a'}$); the CH_3 groups are diastereotopic; the relative assignment of CO-d,d' and CO-e,e' was only made tentatively on the basis of dynamic behavior observed at -30 °C; various spectra between -80 and $+35$ °C depicted [3] IR (hexane): 1921, 1964, 1979, 1994, 2004, 2020, 2046, 2092 (all CO); the IR spectra in pentane solution at -125 °C and in solid state (KBr) are similar indicating similar structures for solid and solution state [3] FD mass spectrum (no medium given): $[M]^+$ [3]
*6 $Os_3(CO)_{10}\{\eta^2$-(neo-C_5H_{11}N=CH)$_2\}$	II (30%) [1, 3] red crystals (from hexane/ether at -80 °C); stable in solution at room temperature [3] ^1H NMR (C_6D_6): 0.81 (s, 6 CH_3), 3.53, 4.10˙(d's, each 2 H of CH_2), 7.36 (s, N=CH); (CD_2Cl_2, -80 °C): 0.88 (s, 6 CH_3), 3.78, 3.90 (d's, each 2 H of CH_2), 7.86 (s, N=CH); all $^2J=12$; the CH_2 groups are diastereotopic [3] ^{13}C NMR (CD_2Cl_2, -80 °C): 27.0, 32.7 (both CH_3), 75.9 (CH_2), 168.6, 173.3, 183.1, 184.0, 189.6 (each 2 CO); no resonance for N=CH given; the CH_3 groups are diastereotopic [3] IR (hexane): 1917, 1964, 1982, 1999, 2013, 2024, 2053, 2099 (all CO); the IR spectra in pentane solution at -125 °C and in solid (KBr) are similar indicating similar structures for solid and solution state [3]
*Further information:	FD mass spectrum (no medium given): $[M]^+$ [3]

$Os_3(CO)_{10}\{\eta^2$-(i-C_3H_7N=CH)$_2\}$ and $Os_3(CO)_{10}\{\eta^2$-(neo-C_5H_{11}N=CH)$_2\}$ (Table 4, No. 5 and No. 6). Variable-temperature ^1H and ^{13}C NMR experiments for Nos. 5 and 6 indicated at least three fluxional processes depending on the temperature range. For No. 5, between -80 and -70 °C a rocking motion of the η^2-(i-C_3H_7N=CH)$_2$ ligand was observed above the Os_3 plane about one axial and two equatorial coordination sites on the Os(1) center (see Fig. 24). Above -70 °C, a second process was observed involving pairwise bridge-terminal CO interchanges perpendicular to the plane of the Os_3 triangle, while above $+20$ °C various processes occurred whereby all carbonyls scramble. Similar fluxional processes were established for No. 6 [3].

$Os_3(CO)_{10}\{\eta^2$-(i-C_3H_7N=CH)$_2\}$ (No. 5) crystallizes in the triclinic, space group $P\bar{1}-C_i^1$ (No. 2) with a = 10.777(2), b = 13.562(2), c = 8.426(2) Å, $\alpha=98.92(2)°$, $\beta=96.45(2)°$, $\gamma=90.00(2)°$; Z = 2. The structure is shown in **Fig. 24**. The trinuclear 48-electron cluster consists of an isosceles Os$_3$ triangle with a mean bond angle of 59.97°. The N(2) atom of the four-electron donating α,α-N,N diimine chelate ligand is in an axial, the N(1) atom in an equatorial position on the Os(1) center; the ligand displays a characteristic bite angle at Os(1) of 76.6(7)°. The N=C double bonds are slightly elongated in comparison to the free ligand, while the central C(1)-C(2) bond is slightly shortened. The plane defined by Os(1), N(1), C(1), C(2), N(2) is almost flat; only N(2) deviates by 0.03 Å out of the least-square plane. The axial CO groups show little tendency for taking semibridging positions leaning toward neighboring Os centers as indicated by rather acute angles. The structure of No. 5 exhibits C$_1$ symmetry in the solid state [3].

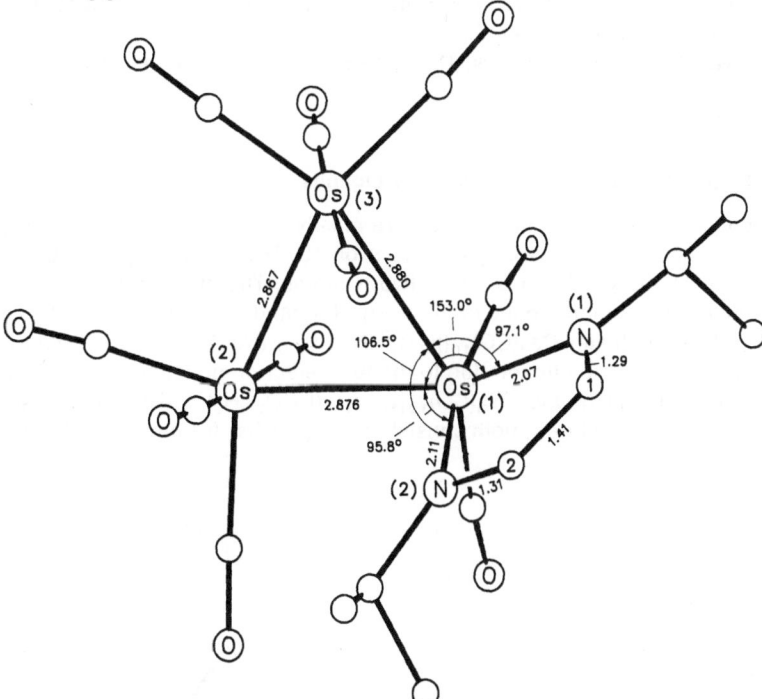

Fig. 24. Molecular structure of Os$_3$(CO)$_{10}${η^2-(i-C$_3$H$_7$N=CH)$_2$} (No. 5) with selected bond distances (in Å) and bond angles [3].

Thermolysis of No. 5 in refluxing hexane for 24 h, or in refluxing heptane for 8 h gave Os$_3${μ$_2$-η^4-(CHNC$_3$H$_7$-i)$_2$}(CO)$_9$ (Formula III); the reaction proceeded via Os$_3${μ$_2$-η^3-CH(NC$_3$H$_7$-i)CH=NC$_3$H$_7$-i}(CO)$_{10}$ (see "Organoosmium Compounds" B 5, 1994, p. 269),

III

based on FT-IR spectra at various temperatures. Similar thermolysis of No. 6 led to a complex mixture from which no pure compound could be isolated [4].

References:

[1] Zoet, R.; van Koten, G.; Vrieze, K.; Duisenberg, A. J. M.; Spek, A. L. (Inorg. Chim. Acta **148** [1988] 71/84).

[2] Zoet, R.; van Koten, G.; Vrieze, K.; Jansen, J.; Goubitz, K.; Stam, C. H. (Organometallics **7** [1988] 1565/72).

[3] Zoet, R.; Jastrzebski, J. T. B. H.; van Koten, G.; Mahabiersing, T.; Vrieze, K.; Heijdenrijk, D.; Stam, C. H. (Organometallics **7** [1988] 2108/17).

[4] Zoet, R.; van Koten, G.; Stufkens, D. J.; Vrieze, K.; Stam, C. H. (Organometallics **7** [1988] 2118/23).

3.1.1.7.3.2 Compounds with μ_2-η^2-Bonded N Ligands

The compounds dealt with in this section are mostly of the type $(\mu\text{-H})Os_3(CO)_n(\mu_2\text{-}\eta^2\text{-L})$ consisting of an isosceles Os_3 triangle; n is 9 or 10. The bridging μ_2-η^2-L ligands are either bonded by N and O, N and S, or by two N atoms. The structures of the compounds showing the different bonding modes of the μ_2-η^2-L ligands are represented by Formulas I to VI. The 48-electron cluster $Os_3(CO)_{10}(\mu_2\text{-}\eta^2\text{-NC}_4H_4N)$ (No. 28) corresponding to Formula IV, has no bridging hydride ligand due to the four-electron donating $C_4H_4N_2$ ligand. Only $Os_3(CO)_{11}\{\mu_2\text{-}\eta^2\text{-N}(C_6F_5)O\}$ and $Os_3(CO)_{12}\{\mu_2\text{-}\eta^2\text{-}(NCO_2CH_3)_2\}$ (Nos. 1 and 40) deviate completely from these structures; both consist of an open Os_3 framework with a nonbonded Os–Os separation.

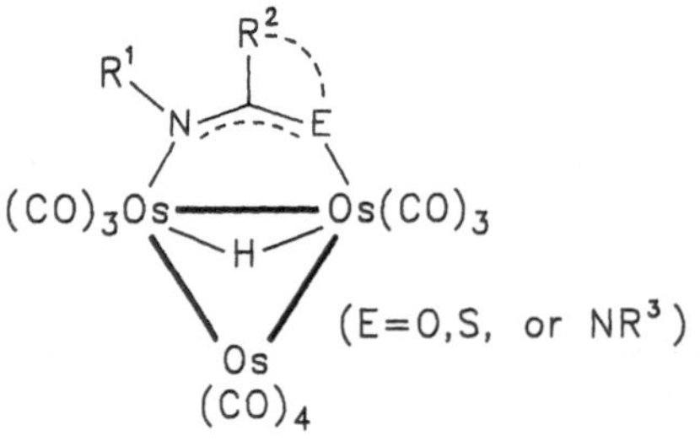

I

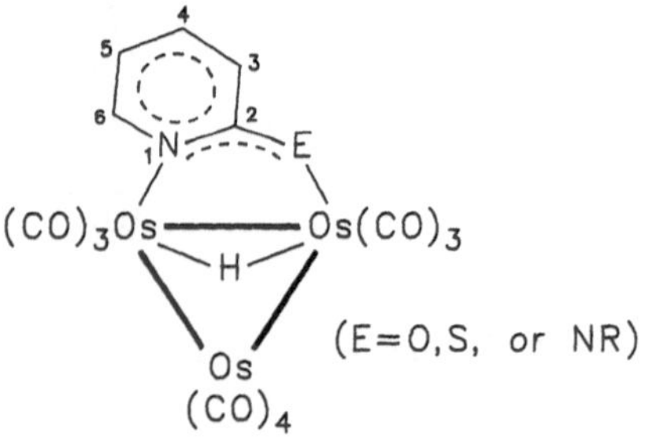

II

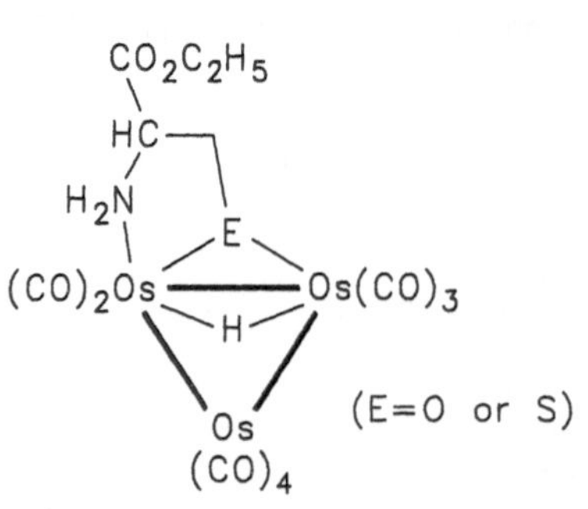

III

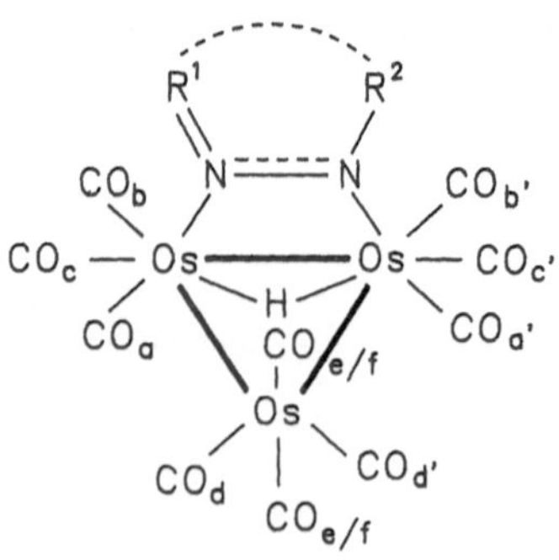

IV

References on p. 90

$$
\begin{array}{c}
\text{C}_6\text{H}_4\text{R}-4 \\
\text{N}=\text{N} \\
(\text{CO})_3\text{Os} \underline{\quad\quad} \text{Os(CO)}_3 \\
\text{H} \\
\text{Os} \\
(\text{CO})_4
\end{array}
\qquad
\begin{array}{c}
\text{H} \qquad\qquad \text{R} \\
\text{N}-\text{N}=\text{N} \\
(\text{CO})_3\text{Os} \underline{\quad\quad} \text{Os(CO)}_3 \\
\text{H} \\
\text{Os} \\
(\text{CO})_4
\end{array}
$$

V   VI

The 48-electron clusters of the type $(\mu\text{-H})\text{Os}_3(\text{CO})_9\{\mu_2\text{-}\eta^2\text{-NH}_2\text{CH}(\text{CO}_2\text{C}_2\text{H}_5)\text{CH}_2\text{E}\}$ (Formula III, E=O, S) were obtained as diastereomeric pairs resulting from a chiralization of the Os_3 metallocycle caused by the tridentate ligand coordination. $(\mu\text{-H})\text{Os}_3(\text{CO})_9\{\mu_2\text{-}\eta^2\text{-}$ $\text{NH}_2\text{CH}(\text{CO}_2\text{C}_2\text{H}_5)\text{CH}_2\text{S}\}$ (Nos. 18 and 19) were obtained in equal amounts, whereas in the case of $(\mu\text{-H})\text{Os}_3(\text{CO})_9\{\mu_2\text{-}\eta^2\text{-NH}_2\text{CH}(\text{CO}_2\text{C}_2\text{H}_5)\text{CH}_2\text{O}\}$ (Nos. 12 and 13) one of the two isomers was formed with a significant predominance. The structures of Nos. 12, 13, and 19 (see Figures 27 to 29) exhibit a reduction of the unbridged Os(1)–Os(3) bond edge which is attributed to the trans effect of the amino group. Nos. 12 and 19 are isostructural with the five-membered chelate rings in an envelope conformation, while the chelate ring in No. 13 is twisted. Furthermore, the ester groups have pseudoaxial orientation in Nos. 12 and 19, but equatorial orientation in No. 13 relative to Os_3. The racemic nature of the monocrystals of Nos. 12, 13, and 19 presupposes the presence of an amino acid ligand with D configuration; the most likely reason for the formation of D– from L–amino acids is probably caused by racemization of the L forms upon tridentate coordination at the Os_3 metallocycle [24].

Most compounds listed in Table 5 undergo decarbonylation and/or rearrangement of the bridging ligand upon thermolysis in refluxing hydrocarbons in a sealed tube. The compounds of type I mainly rearranged by loss of one CO group to give clusters with $\mu_3\text{-}\eta^1\text{-}$bridging ligands (Section 3.1.1.7.2.4), $\mu_3\text{-}\eta^2$-bridging ligands (Section 3.1.1.7.3.3), $\mu_2\text{-}\eta^2\text{-}$, and $\mu_2\text{-}\eta^3$-bridging ligands bonded by C and a heteroatom (see "Organoosmium Compounds" B 5, 1994, pp. 145, 146, and 264). In many thermolysis reactions $\text{Os}_6(\text{CO})_{18}$ (see "Organoosmium Compounds" B 9, 1995, Section 6.1) was formed exclusively or in addition to the above mentioned complexes [3, 6, 10, 15, 20]. The compounds of Formulas II and VI mainly decarbonylated giving clusters with $\mu_3\text{-}\eta^2$- or $\mu_3\text{-}\eta^1$-bridging ligands (Sections 3.1.1.7.3.3 and 3.1.1.7.2.4), respectively [3, 7 to 9, 11].

Most of the compounds listed in Table 5 were prepared by the following methods:

Method I: Compounds of the type $(\mu\text{-H})\text{Os}_3(\text{CO})_{10}(\mu_2\text{-}\eta^2\text{-NR}^1\text{CR}^2\text{=E})$ (Formula I, E=O or NR^3) were prepared from $\text{Os}_3(\text{CO})_{10}(\text{NCCH}_3)_2$

                a. by reaction with an excess of $\text{NH}_2\text{CR}^2\text{=O}$ (R^2=H, CH_3, C_2H_5, n-C_3H_7, C_6H_5) in refluxing cyclohexane for 2.5 to 4 h and workup by TLC with CH_2Cl_2/hexane (1:4) as eluant [5, 20], or

                b. by treatment with $\text{NH}_2\text{CR}^2\text{=NR}^3$ (ca. 1:1) in benzene at room temperature for 12 h and workup as before; R^2=CH_3, C_6H_5; R^3=H, C_6H_5 [15].

Method II: Compounds of the type $(\mu\text{-H})\text{Os}_3(\text{CO})_{10}(\mu_2\text{-}\eta^2\text{-NHN=NR})$ (Formula VI) were prepared from $(\mu\text{-H})_2\text{Os}_3(\text{CO})_{10}$ and an excess of the corresponding azide RN_3 in benzene [21] or in hexane at 20 °C for 24 h, or 3 weeks as in the case of $\text{C}_6\text{H}_5\text{N}_3$ [8, 9, 12]; workup by TLC with CH_2Cl_2/hexane (2:3) as eluant [8, 9, 12, 21]. In the case of HN_3, the reaction mixture was extracted with 5% HCl several times to remove excess HN_3 [21].

References on p. 90

$(\mu\text{-}H)Os_3(CO)_{10}\{\mu_2\text{-}\eta^2\text{-}NHC(C_6H_5){=}O\}$ (No. 6) and $(\mu\text{-}H)Os_3(CO)_{10}\{\mu_2\text{-}\eta^2\text{-}NHC(NHC_6H_4){=}N\}$ (No. 24; see also Formula I) were prepared similarly from $C_6H_5C(O)N_3$ or $\{cyclo\text{-}C_6H_4NHC(N)\}N_3$ in refluxing hexane; these compounds were probably formed by loss of N_2 from an intermediate azide adduct [9].

Method III: Compounds of the type $(\mu\text{-}H)Os_3(CO)_{10}(\mu_2\text{-}\eta^2\text{-}NR^1CH{=}E)$ (Formula I, $E{=}O$ or S) were prepared from $(\mu\text{-}H)_2Os_3(CO)_{10}$ and an excess of $NR^1{=}C{=}E$ neat at room temperature for 40 min [6, 18], or in refluxing hexane for more than 3 h [2, 4]; workup by column chromatography with CH_2Cl_2/hexane (1:4) as eluant [6, 18] or with hexane/benzene (1:1) over $Al_2O_3 \cdot 6\%$ H_2O [2].

Method IV: Compounds of the type $(\mu\text{-}H)Os_3(CO)_{10}(\mu_2\text{-}\eta^2\text{-}N{=}NC_6H_4R\text{-}4)$ (Formula V) were prepared from $(\mu\text{-}H)Os_3(CO)_{10}(\mu_2\text{-}\eta^1\text{-}N{=}NC_6H_4R\text{-}4)$ (R=H, F, CH_3; Section 3.1.1.7.2.3) by UV irradiation (450-W medium-pressure mercury-vapor lamp) in heptane for 4 to 6 h under 1 atm of CO; workup by TLC with petroleum ether as eluant [19].

Table 5
Compounds with $\mu_2\text{-}\eta^2$-Bonded N Ligands.
An asterisk preceding the compound number indicates further information at the end of the table, pp. 81/90.
Explanations, abbreviations, and units on p. X.

No. compound	method of preparation (yield) properties and remarks

compounds with $\mu_2\text{-}\eta^2$-bridging ligands bonded by N and O atoms

*1 $Os_3(CO)_{11}\{\mu_2\text{-}\eta^2\text{-}N(C_6F_5)O\}$	from $Os_3(CO)_{11}NCCH_3$ (see "Organoosmium Compounds" B 3, 1994, p. 237) by dropwise addition of an excess of C_6F_5NO in THF at room temperature over 20 min; workup by TLC with CH_2Cl_2/hexane (1:4) as eluant (14%) [25] yellow crystals (from CH_2Cl_2/hexane) [25] ^{19}F NMR ($CDCl_3$, vs. CF_3CO_2H): -84.90 (m, 2 F), -81.26 (m, 1 F), -67.94 (m, 2 F) [25] IR (no medium given): 1006 (probably νNO), 1513 (C_6F_5 ring vibration), 1976 sh, 1982 s, 1997 s, 2006 vs, 2024 vs, 2030 sh, 2041 m, 2052 vs, 2057 vs, 2062 vs, 2075 m, 2087 vs, 2106 w, 2113 w, 2134 w (all CO) [25] the HPLC separation from $Os_3(CO)_9(\mu_3\text{-}\eta^1\text{-}NC_6F_5)_2$ (Section 3.1.1.7.2.4), $Os_3(CO)_{11}NCCH_3$, and $Os_3(CO)_{12}$ was investigated [25]
2 $(\mu\text{-}H)Os_3(CO)_{10}(\mu_2\text{-}\eta^2\text{-}NHCH{=}O)$ Formula I	Ia (50%), along with $(\mu\text{-}H)_2Os_3(CO)_{10}$ [5, 20] yellow microcrystalline solid (from hexane) [20] ^1H NMR (CD_2Cl_2): -11.84 (s, OsH), 5.59 (br, NH), 7.93 (d, CH) [20]

Table 5 (continued)

No. compound	method of preparation (yield) properties and remarks
	IR (cyclohexane): 1597 (NHCO), 1979 mw, 1985 mw, 1988 sh, 2001 m, 2013 s, 2024 s, 2059 s, 2070 s, 2110 mw (all CO) [5, 20] mass spectrum: [M]$^+$ [5, 20] thermolysis in refluxing nonane for 40 h [20] gave (μ-H)Os$_3$(CO)$_{10}$(μ_2-η^1-NH$_2$) (Section 3.1.1.7.2.1) in 50% yield [20]; earlier the formation of a compound of the type **(μ-H)Os$_3$(CO)$_9$(μ_3-η^2-NHCH=O)** was assumed [5]; see also p. 92
3 (μ-H)Os$_3$(CO)$_{10}${μ_2-η^2-NHC(CH$_3$)=O} Formula I	Ia (75%) [5, 20] greenish-yellow crystals (from pentane) [20] ^{1}H NMR (CD$_2$Cl$_2$): −11.90 (s, OsH), 1.94 (s, CH$_3$), 5.29 (NH) [20] IR (cyclohexane): 1576 (NHCO), 1977 mw, 1983 mw, 1986 sh, 1999 m, 2009 s, 2023 s, 2057 s, 2069 s, 2109 mw (all CO) [5, 20] mass spectrum: [M]$^+$ [5, 20] thermolysis in refluxing nonane for 24 h gave Os$_6$(CO)$_{18}$ (see "Organoosmium Compounds" B 9, 1995, Section 6.1) in 75% yield [20]
4 (μ-H)Os$_3$(CO)$_{10}${μ_2-η^2-NHC(C$_2$H$_5$)=O} Formula I	Ia (70%), along with 10% of Os$_3$(μ_2-η^2-O=CC$_2$H$_5$)(CO)$_{10}$(μ-NH$_2$) (see "Organoosmium Compounds" B 5, 1994, p. 145) [5, 20] yellow crystals (from pentane) [20] ^{1}H NMR (CD$_2$Cl$_2$): −11.80 (s, OsH), 0.97 (t, CH$_3$), 2.19 (q, CH$_2$); no resonance for NH given [20] IR (cyclohexane): 1570 (NHCO), 1977 mw, 1984 mw, 1987 sh, 1999 m, 2011 s, 2023 s, 2058 s, 2069 s, 2109 mw (all CO) [5, 20] mass spectrum: [M]$^+$ [5, 20] thermolysis in refluxing nonane for 24 h gave Os$_6$(CO)$_{18}$ (30%) and Os$_3$(μ_2-η^2-O=CC$_2$H$_5$)(CO)$_{10}$(μ-NH$_2$) (10%) [20]
5 (μ-H)Os$_3$(CO)$_{10}${μ_2-η^2-NHC(C$_3$H$_7$-n)=O} Formula I	Ia (70%), along with 10% of Os$_3$(μ_2-η^2-O=CC$_3$H$_7$-n)(CO)$_{10}$(μ-NH$_2$) (see "Organoosmium Compounds" B 5, 1994, p. 146) [5, 20]

References on p. 90

Table 5 (continued)

No. compound	method of preparation (yield) properties and remarks
5 (continued)	yellow crystals (from hexane) [20] ^{1}H NMR (CD_2Cl_2): -11.76 (s, OsH), 0.85 (t, CH_3; $^3J=7$), 1.52, 2.20 (m's, both CH_2); no resonance for NH given [20] IR (cyclohexane): 1570 (NHCO), 1976 mw, 1982 mw, 1985 sh, 1998 m, 2010 s, 2022 s, 2056 s, 2068 s, 2108 mw (all CO) [5, 20] mass spectrum: $[M]^+$ [5, 20] thermolysis in refluxing nonane for 24 h gave $Os_6(CO)_{18}$ (30%) and $Os_3(\mu_2-\eta^2-O=CC_3H_7-n)(CO)_{10}(\mu-NH_2)$ (10%) [20]
6 $(\mu-H)Os_3(CO)_{10}\{\mu_2-\eta^2-NHC(C_6H_5)=O\}$ Formula I	Ia (70%), along with 10% of $Os_3(\mu_2-\eta^2-O=CC_6H_5)(CO)_{10}(\mu-NH_2)$ (see "Organoosmium Compounds" B 5, 1994, p. 146) and $Os_2(CO)_6\{\mu-\eta^2-O=C(C_6H_5)NH\}_2$ [5, 20] II (21%) [9] yellow crystals (from hexane) [20] ^{1}H NMR (CD_2Cl_2): -11.60 (s, OsH), 6.06 (br, NH), 7.42 (s, C_6H_5) [20] IR (cyclohexane): 1590 (NHCO), 1977 mw, 1984 mw, 1987 sh, 1999 m, 2012 s, 2023 s, 2059 s, 2070 s, 2110 m (all CO) [5, 20] mass spectrum: $[M]^+$ [5, 20] thermolysis in refluxing nonane for 45 h yielded $Os_6(CO)_{18}$ (30%), and $Os_3(\mu_2-\eta^2-O=CC_6H_5)(CO)_{10}(\mu-NH_2)$ (10%; "Organoosmium Compounds" B 5, 1994, p. 146); a third compound was tentatively assigned to the structure $H_4Os_4(C_6H_5C_2C_6H_5)(CO)_{11}$ (see "Organoosmium Compounds" B 8, 1995, Section 4.7), based on IR and mass spectroscopic data [20]
7 $(\mu-H)Os_3(CO)_{10}\{\mu_2-\eta^2-N(CH_3)CH=O\}$ Formula I	III (80%), along with 6% of $(\mu-H)Os_3\{\mu_2-\eta^2-C(OH)=NCH_3\}(CO)_{10}$ and 10% of $(\mu-H)Os_3-\{\mu_2-\eta^2-C(O_2CNHCH_3)=NCH_3\}(CO)_{10}$ (both "Organoosmium Compounds" B 5, 1994, pp. 194/5) [6, 18] ^{1}H NMR ($CDCl_3$): -11.46 (s, OsH), 3.27 (s, CH_3), 7.51 (s, CH) [18] IR (C_2Cl_4): 1450 (δCH_3), 1601 ($\nu NCHO$), 2912 (νCH_3); (hexane): 1980 w, 1988 w(sh), 1989 w, 2000 m, 2012 s(sh), 2013 s, 2025 vs, 2059 s, 2070 s, 2109 m (all CO) [18]

References on p. 90

Table 5 (continued)

No. compound	method of preparation (yield) properties and remarks

decarbonylated above 150 °C giving
$(\mu\text{-H})_2Os_3(CO)_9(\mu_3\text{-}\eta^1\text{-NCH}_3)$
(Section 3.1.1.7.2.4) [6]

8 $(\mu\text{-H})Os_3(CO)_{10}\{\mu_2\text{-}\eta^2\text{-N}(C_2H_5)CH\text{=}O\}$
Formula I
III (80%), along with 6% $(\mu\text{-H})Os_3\{\mu_2\text{-}\eta^2\text{-}$
$C(OH)\text{=}NC_2H_5\}(CO)_{10}$ and 10% of $(\mu\text{-H})Os_3\text{-}$
$\{\mu_2\text{-}\eta^2\text{-C}(O_2CNHC_2H_5)\text{=}NC_2H_5\}(CO)_{10}$ (both
"Organoosmium Compounds" B 5, 1994,
pp. 195/6) [18]
^{1}H NMR (CDCl$_3$): -11.44 (s, OsH), 1.14 (t, CH$_3$;
$^3J\text{=}7.0$), 3.50 (m, CH$_2$), 7.57 (s, CH) [18]
IR: similar to that of No. 7 [18]

*9 $(\mu\text{-H})Os_3(CO)_{10}\{\mu_2\text{-}\eta^2\text{-N}(C_6H_4CH_3\text{-}4)CH\text{=}O\}$
Formula I
III (68%), along with 1% of $(\mu\text{-H})Os_3(\mu_2\text{-}\eta^2\text{-}$
$O\text{=}CNHC_6H_4CH_3\text{-}4)(CO)_{10}$ (see "Organoosmium Compounds" B 5, 1994, p. 116) and
two other $\mu\text{-H}$ containing complexes, based
on ^{1}H NMR spectra [2, 4]
yellow crystals (from pentane or hexane at
-20 °C), m.p. 128.5 to 130 °C [2, 4]
^{1}H NMR (acetone-d$_6$): -11.35 (s, OsH), 2.34
(CH$_3$), 7.01 (C$_6$H$_4$), 7.65 (CH) [4]; (no medium
given): 7.65 (CH); no other resonances given
[2]
IR (hexane): 1980 m, 1987 m, 2005 s(sh),
2010 s, 2030 s, 2040 w, 2060 s, 2075 s,
2110 m (NH and CO) [2, 4]
reaction with P(CH$_3$)$_2$C$_6$H$_5$ in refluxing heptane
led to $(\mu\text{-H})Os_3(CO)_9P(CH_3)_2C_6H_5\{\mu_2\text{-}\eta^2\text{-}$
$N(C_6H_4CH_3\text{-}4)CH\text{=}O\}$ (see "Organoosmium
Compounds" B 4b, in preparation) in 51%
yield [4]
attempted thermolysis in refluxing heptane
failed [4]

10 $(\mu\text{-H})Os_3(CO)_{10}\{\mu_2\text{-}\eta^2\text{-N}(C_6H_5)CH\text{=}O\}$
Formula I
III [2, 4]
m.p. 142 to 144 °C [4]
^{1}H NMR (acetone-d$_6$): -10.49 (s, OsH), 7.27
(m, C$_6$H$_5$), 7.74 (s, CH) [4]; (no medium
given): 7.74 (CH); no other resonances given
[2]
IR (hexane): 1965 w(sh), 1985 m(sh), 2055 s,
2065 s, 2100 m (NH and CO) [4]

References on p. 90

Table 5 (continued)

No. compound	method of preparation (yield) properties and remarks
11 $(\mu-H)Os_3(CO)_{10}(\mu_2-\eta^2-NC_5H_4O)$ Formula II	from $Os_3(CO)_{10}(NCCH_3)_2$ and 2-hydroxypyridine in refluxing cyclohexane for 1 h (75%) [5], or in benzene at 20 °C for 18 h and workup by TLC with CH_2Cl_2/hexane as eluant (36%) [7]

from $Os_3(CO)_{12}$ and 2-hydroxypyridine in toluene at 145 °C for 80 h under vacuum in a sealed tube, followed by chromatography with pentane/toluene (1:1) as eluant (10%) [11]

also obtained by carbonylation of $(\mu-H)Os_3(CO)_9(\mu_3-\eta^2-OC_5H_4N)$ (Section 3.1.1.7.3.3) [10]

yellow crystals (from hexane at 0 °C) [11]; stable under an Ar atmosphere up to 200 °C [7]

^1H NMR $(CDCl_3)$: -10.73 (s, OsH), 6.23 (dt, H–5), 6.43 (dd, H–3), 7.11 (dt, H–4), 8.08 (dd, H–6); no values for ^3J and ^4J given [11]; similar in CD_2Cl_2 [5]

IR (cyclohexane): 1981 m, 1984 m, 1991 m, 2002 m, 2016 sh, 2018 s, 2026 s, 2061 s, 2071 s, 2112 m (all CO) [11]; similar data [5]

mass spectrum: $[M]^+$ [5]

decarbonylation in refluxing toluene for 4 h gave $(\mu-H)Os_3(CO)_9(\mu_3-\eta^2-OC_5H_4N)$ (Section 3.1.1.7.3.3) [10, 11]; the decarbonylation is reversible [10]

*12 $(\mu-H)Os_3(CO)_9\{\mu_2-\eta^2-NH_2CH(CO_2C_2H_5)CH_2O\}$ Formula III	diastereomeric with No. 13 [24]

from $(\mu-H)Os_3(CO)_{10}(\mu-OH)$ (see "Organo-osmium Compounds" B 3, 1994, p. 94) and ca. 1.5 equivalents of $(CH_3)_3NO \cdot 2 H_2O$ (dissolved in C_2H_5OH) for 1 h, followed by treatment with an excess of L–α–serine ethyl ester in THF at ca. 20 °C under reduced pressure; direct workup by TLC with benzene/hexane/acetone (3:2:1) as eluant (51.8% as inseparable mixture with No. 13); crystallization from CH_2Cl_2/hexane at +10 °C gave single crystals for both compounds which can be separated by hand [24]

yellow orthorhombic prisms [24]

References on p. 90

Table 5 (continued)

No. compound	method of preparation (yield) properties and remarks

¹H NMR (CDCl₃; mixture of Nos. 12 and 13):
     − 11.57, − 10.71 (s's, each 1 OsH), 1.28 (t,
     CH₃), 3.73, 4.23 (m's, OCH₂, CH, NH₂, CH₂ of
     C₂H₅) [24]

^{1}H NMR (CDCl$_3$; mixture of Nos. 12 and 13):
− 11.57, − 10.71 (s's, each 1 OsH), 1.28 (t, CH$_3$), 3.73, 4.23 (m's, OCH$_2$, CH, NH$_2$, CH$_2$ of C$_2$H$_5$) [24]

IR (cyclohexane; mixture of Nos. 12 and 13): 1930 m, 1971 w (both CO$_2$C$_2$H$_5$), 1971 m, 1996 s, 2017 vs, 2055 s, 2100 m (all CO), 3284, 3341 w (both NH) [24]

EI mass spectrum (for Nos. 12 and 13): [M]$^+$ [24]

*13 (μ-H)Os$_3$(CO)$_9${μ$_2$-η^2-NH$_2$CH(CO$_2$C$_2$H$_5$)CH$_2$O}
    Formula III

diastereomeric with No. 12 [24]
for formation, ^{1}H NMR and IR spectra, see No. 12; no other data given [24]
yellow truncated parallelepiped crystals (from CH$_2$Cl$_2$/hexane at + 10 °C) [24]

14 (μ-H)Os$_3$(CO)$_9$(μ$_2$-η^2-NC$_5$H$_4$CH$_2$O)

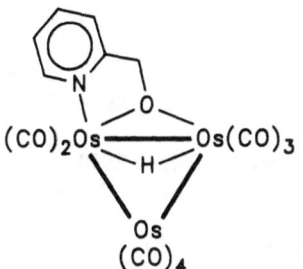

no preparation reported
IR (cyclohexane): 1936, 1975, 1981, 1991, 2016, 2056, 2099 (all CO) [14]

15 (μ-H)Os$_3$(CO)$_9${μ$_2$-η^2-N(CH$_2$C$_6$H$_5$)=CHC$_6$H$_4$O}

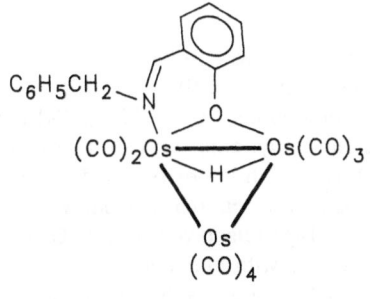

by thermolysis of (μ-H)Os$_3$(μ$_2$-η^2-O=C$_6$H$_3$=CHNHCH$_2$C$_6$H$_5$)(CO)$_{10}$ (see "Organoosmium Compounds" B 5, 1994, p. 163) in refluxing heptane for 2.5 h; workup by TLC with ether/petroleum ether (1:4) as eluant (small amounts), along with (μ-H)Os$_3$-(μ$_3$-η^2-O=C$_6$H$_3$CH=NHCH$_2$C$_6$H$_5$)(CO)$_9$ (see "Organoosmium Compounds" B 6, 1993, p. 76) and (μ-H)$_2$Os$_3$(μ$_3$-η^2-OC$_6$H$_2$CHNHCH$_2$C$_6$H$_5$)(CO)$_9$ as the main product [14]
yellow crystals [14]
^{1}H NMR (CDCl$_3$): − 10.77 (s, OsH), 5.42, 5.50 (d's, each 1 H of CH$_2$; ^{2}J = 13.4), 7.0 to 7.5 (m, C$_6$H$_4$), 8.19 (s, CHN) [14]
IR (cyclohexane): 1939 m, 1975 m, 1984 m, 1991 m, 2002 s, 2013 vs, 2056 s, 2096 m (all CO) [14]

References on p. 90

Table 5 (continued)

No. compound	method of preparation (yield) properties and remarks

compounds with μ_2-η^2-bridging ligands bonded by N and S atoms

16 $(\mu$-H)Os$_3$(CO)$_{10}$\{μ_2-η^2-N(C$_6$H$_5$)CH=S\}
Formula I

III [2]
^{1}H NMR (no medium given): 8.12 (CH); no other resonances given [2]

17 $(\mu$-H)Os$_3$(CO)$_{10}$(μ_2-η^2-NC$_5$H$_4$S)
Formula II

from Os$_3$(CO)$_{10}$(NCCH$_3$)$_2$ and pyridine-2-thiol (ca. 1:1) in benzene; workup by TLC with CH$_2$Cl$_2$/hexane as eluant (62%) [7]
from Os$_3$(CO)$_{12}$ and pyridine-2-thiol (ca. 1:3) in toluene at reflux temperature for 11 h; workup by TLC with petroleum ether as eluant (49%), along with 19% of $(\mu$-H)Os$_3$-(CO)$_9$(μ_3-η^2-SC$_5$H$_4$N) (Section 3.1.1.7.3.3) and 30% of Os(CO)$_2$(SC$_5$H$_4$N)$_2$ (see "Organoosmium Compounds" A 2, 1993, p. 22) [23]
yellow solid [23]
^{1}H NMR (no medium given): -17.09 (s, OsH), 7.23 (d, 1 H; J=3), 7.46 to 7.83 (m, 2 H), 8.54 (d, 1 H; J=3) [7]
IR (no medium given): 1983 w, 2000 m(d), 2014 m(d), 2021 s, 2058 m(d), 2068 s, 2108 w [7]
decarbonylation in heptane at 160 °C yielded $(\mu$-H)Os$_3$(CO)$_9$(μ_3-η^2-SC$_5$H$_4$N) (Section 3.1.1.7.3.3) in 67% yield [7]

18 $(\mu$-H)Os$_3$(CO)$_9$\{μ_2-η^2-NH$_2$CH(CO$_2$C$_2$H$_5$)CH$_2$S\}
Formula III

diastereomeric with No. 19 [26]
from $(\mu$-H)Os$_3$(CO)$_{10}$(μ-OH) (see "Organoosmium Compounds" B 3, 1994, p. 94) and L-α-cysteine ethyl ester (1:1 or 1:2) in THF at ca. 20 °C for 1 h under reduced pressure in the presence of ca. 0.5 to 1 equivalent of (CH$_3$)$_3$NO·2 H$_2$O (dissolved in C$_2$H$_5$OH); workup by TLC with benzene/hexane (2:3) as eluant (29%), along with 30.5% of No. 19 [24, 26]
by treatment of $(\mu$-H)Os$_3$(CO)$_{10}$\{μ-SCH$_2$-CH(NH$_2$)CO$_2$C$_2$H$_5$\} with (CH$_3$)$_3$NO·2 H$_2$O (dissolved in CH$_3$OH) in THF at ca. 20 °C for 1.5 h; separation as before (20.5%), along with 22.4% of No. 19 [26]
$\alpha_D^{22} = -191.49°$ in CHCl$_3$ (0.528 M) [24, 26]

References on p. 90

Table 5 (continued)

No. compound	method of preparation (yield) properties and remarks
	¹H NMR (CD₂Cl₂): − 16.07 (s, OsH), 1.26 (t, CH₃), 3.51 (m, SCH₂CH), 4.21 (q, CH₂ of C₂H₅), 4.85 (br, NH₂) [24, 26]; spectrum depicted [26]

^{1}H NMR (CD$_2$Cl$_2$): -16.07 (s, OsH), 1.26 (t, CH$_3$), 3.51 (m, SCH$_2$CH), 4.21 (q, CH$_2$ of C$_2$H$_5$), 4.85 (br, NH$_2$) [24, 26]; spectrum depicted [26]

IR (cyclohexane): 1739 w (CO$_2$C$_2$H$_5$), 1931 m, 1972 m, 1977 sh, 1995 s, 2015 vs, 2054 s, 2097 m (all CO); (CCl$_4$): 3275 w, 3336 w (both NH) [24, 26]

EI mass spectrum: [M]$^+$, [M$-$n CO]$^+$, n$=$1 to 9 [24, 26]

attempted carbonylation of Nos. 18 and 19 in benzene at 80 °C for 24 h failed [26]

*19 (μ-H)Os$_3$(CO)$_9${μ_2-η^2-NH$_2$CH(CO$_2$C$_2$H$_5$)CH$_2$S}
    Formula III

diastereomeric with No. 18 [24, 26]

for formation, see No. 18 [24, 26]

yellow crystals (from CH$_2$Cl$_2$/hexane at $+10$ °C) [24]

$\alpha_D^{22} = +142.43°$ in CHCl$_3$ (0.355 M) [24, 26]

^{1}H NMR (CD$_2$Cl$_2$): -16.57 (s, OsH), 1.26 (t, CH$_3$), 2.36 (m, SCH$_2$), 3.01 (m, CH), 4.23 (q, CH$_2$ of C$_2$H$_5$), 4.50 (br, NH$_2$) [24, 26]; spectrum depicted [26]

IR (cyclohexane): 1740 w (CO$_2$C$_2$H$_5$), 1930 m, 1969 m, 1993 s, 2013 vs, 2052 s, 2096 m (all CO); (CCl$_4$): 3288 w, 3335 w (both NH) [24, 26]

EI mass spectrum: [M]$^+$, [M$-$n CO]$^+$, n$=$1 to 9 [24, 26]

for carbonylation, see No. 18 [26]

*20 Os$_3$(CO)$_9$(μ_2-η^2-N$_2$C$_6$H$_7$S)(μ-SC$_6$H$_7$N$_2$)

by isomerization of Os$_3$(CO)$_9$(μ_3-η^2-N$_2$C$_6$H$_7$S)-(μ-SC$_6$H$_7$N$_2$) (Section 3.1.1.7.3.3) in CH$_2$Cl$_2$/h-heptane (1:1) at 25 °C for 10 d; workup by TLC with CH$_2$Cl$_2$/hexane (1:4) as eluant (87%) [27]

pale orange crystals (from CH$_2$Cl$_2$) [27]

^{1}H NMR (CD$_2$Cl$_2$?): 2.29, 2.35 (s's, each 1 CH$_3$), 2.32 (s, 2 CH$_3$), 6.58, 6.63 (s's, each 1 CH); no further assignments given [27]

IR (CH$_2$Cl$_2$): 1922 m, 1989 s, 2006 vs, 2059 s, 2096 s (all CO) [27]

FAB mass spectrum: [M]$^+$ [27]

References on p. 90

Table 5 (continued)

No. compound	method of preparation (yield) properties and remarks

compounds with μ_2-η^2-bridging ligands bonded by two N atoms

21 $(\mu$-H)Os$_3$(CO)$_{10}${μ_2-η^2-NHC(CH$_3$)=NH}
Formula I

Ib (48%) [15]
^{1}H NMR (CDCl$_3$): -13.36 (s, OsH), 2.00 (s, CH$_3$), 3.94 (br s, NH) [15]
IR (hexane): 1973 w, 1984 m, 1994 m, 2006 s, 2021 s, 2055 s, 2065 s, 2107 w (all CO) [15]
mass spectrum: [M]$^+$ [15]
decarbonylation in hexane at 140 °C gave two tautomeric forms of $(\mu$-H)Os$_3$(CO)$_9$(μ_3-η^2-C$_2$H$_5$N$_2$) (Section 3.1.1.7.3.3) in nearly equal amounts [15]

22 $(\mu$-H)Os$_3$(CO)$_{10}${μ_2-η^2-NHC(C$_6$H$_5$)=NC$_6$H$_5$}
Formula I

Ib (52%) [15]
^{1}H NMR (acetone-d$_6$): -12.49 (s, OsH), 6.27 (br s, NH), 6.73 to 7.29 (m, C$_6$H$_5$) [15]
IR (hexane): 1972 m, 1982 m, 1994 m, 2003 s, 2020 s, 2051 s, 2065 s, 2105 w (all CO) [15]
mass spectrum: [M]$^+$ [15]
decarbonylation in heptane at 97 °C for 6 h gave $(\mu$-H)Os$_3$(CO)$_9${μ_3-η^2-NHC(C$_6$H$_5$)=NC$_6$H$_5$} (Section 3.1.1.7.3.3) [15]

23 $(\mu$-H)Os$_3$(CO)$_{10}${μ_2-η^2-N(CH$_2$C$_6$H$_5$)CH=NCH$_2$C$_6$H$_5$}
Formula I

from Os$_3$(C$_8$H$_{14}$)$_2$(CO)$_{10}$ (dissolved in cyclo-octene) and C$_6$H$_5$CH$_2$NHCH=NCH$_2$C$_6$H$_5$ in CH$_2$Cl$_2$ for 80 min; workup by TLC with pentane as eluant (15%), along with unchar-acterized compounds [3, 10]
by carbonylation of $(\mu$-H)$_2$Os$_3$(μ_2-η^3-C$_6$H$_4$CH$_2$NCHNCH$_2$C$_6$H$_5$)(CO)$_9$ (see "Organo-osmium Compounds" B 5, 1994, p. 264) in refluxing octane; similar workup as before (37%) [3, 10]
yellow crystals (from hexane at 0 °C) [10]
^{1}H NMR (CDCl$_3$, 29 °C): -12.91 (d, OsH; J=2), 3.26, 3.66 (d's, both CH$_2$; ^{2}J=14), 7.14 to 7.29 (m, C$_6$H$_5$), 7.87 (d, CH; ^{4}J=2) [3, 10]; the two CH$_2$C$_6$H$_5$ groups are equivalent, their CH$_2$ protons are diastereotopic [10]
IR (cyclohexane): 1975 mw, 1986 mw, 1995 m, 2004 s, 2022 s, 2055 s, 2065 s, 2106 m (all CO) [3, 10]
decarbonylation in refluxing octane led to $(\mu$-H)$_2$Os$_3$(μ_2-η^3-C$_6$H$_4$CH$_2$NCHNCH$_2$C$_6$H$_5$)-(CO)$_9$ (see "Organoosmium Compounds"

Table 5 (continued)

No. compound	method of preparation (yield) properties and remarks

B 5, 1994, p. 264) by ortho-metallation of one of the benzyl groups, while the bridging formamidinato ligand remains essentially unchanged [3, 10]

24 $(\mu\text{-H})Os_3(CO)_{10}\{\mu_2\text{-}\eta^2\text{-NHC}(NHC_6H_4)=N\}$
Formula I

II (25%) [9]
^{1}H NMR (acetone-d_6): -15.04 (s, OsH), 3.36, 9.33 (br, both NH), 6.89 (m, C_6H_4) [9]
IR (hexane): 1988 m, 1994 sh, 2010 s, 2016 s, 2052 s, 2060 s, 2102 w (all CO) [9]
EI mass spectrum: $[M]^+$ [9]

25 $(\mu\text{-H})Os_3(CO)_{10}(\mu_2\text{-}\eta^2\text{-NC}_5H_4NH)$
Formula II

from $Os_3(C_8H_{14})_2(CO)_{10}$ (dissolved in cyclooctene) and 2-aminopyridine at room temperature for 88 h; workup by TLC (65%) [11]
from $Os_3(CO)_{10}(NCCH_3)_2$ and 2-aminopryidine (ca. 1:1) in benzene; workup by TLC with CH_2Cl_2/hexane as eluant (35%) [7]
by carbonylation of $(\mu\text{-H})Os_3(CO)_9(\mu_3\text{-}\eta^2\text{-}NHC_5H_4N)$ (Section 3.1.1.7.3.3) [10]
yellow crystals [11]
^{1}H NMR (CDCl$_3$, 27 °C): -12.24 (s, OsH), 3.85 (NH); 5.85 (dt, H–5), 6.12 (dd, H–3), 6.72 (dt, H–4), 7.87 (dd, H–6) [11]
IR (cyclohexane): 1975 m, 1985 m, 1988 m, 1996 m, 2005 m, 2014 s, 2022 s, 2058 s, 2065 s, 2108 m (all CO) [11]
thermolysis in heptane at 68 °C for 5 h [7] or in cyclohexane at 81 °C for 8 h [11] gave $(\mu\text{-H})Os_3(CO)_9(\mu_3\text{-}\eta^2\text{-}NHC_5H_4N)$ (Section 3.1.1.7.3.3) [7, 11]

26 $(\mu\text{-H})Os_3(CO)_{10}(\mu_2\text{-}\eta^2\text{-NC}_5H_4NCH_2C_6H_5)$
Formula II

from $Os_3(C_8H_{14})_2(CO)_{10}$ (dissolved in cyclooctene) and N–benzyl-2-aminopyridine in CH_2Cl_2 at room temperature for 20 h, followed by workup by TLC (33%) [3, 11]
yellow crystals (from hexane) [11]
^{1}H NMR (CDCl$_3$, 29 °C): -11.72 (s, OsH), 4.54, 5.02 (d's, each 1 H of CH_2; $^2J=17$), 5.95 (dt, H–5), 6.24 (dd, H–3), 6.89 (dt, H–4), 8.18 (dd, H–6), 7.0 to 7.4 (C_6H_5); the methylene protons showed an AB quartet since the molecule is chiral having the amidopyridine ligand lying above the Os_3 plane; there is no rapid enantiomerization leading to coalescence [3, 11]

References on p. 90

Table 5 (continued)

No. compound	method of preparation (yield) properties and remarks
26 (continued)	IR (cyclohexane): 1975 m, 1986 m, 1995 m, 2004 m, 2012 s, 2022 s, 2057 s, 2064 s, 2107 m (all CO) [3, 11] decarbonylation in refluxing heptane for 45 min led to a mixture of isomeric $(\mu-H)Os_3(CO)_9\{\mu_3-\eta^2-N(CH_2C_6H_5)C_5H_4N\}$ (Section 3.1.1.7.3.3) and $(\mu-H)_2Os_3(\mu_2-\eta^3-C_6H_4CH_2NC_5H_4N)(CO)_9$ (see "Organoosmium Compounds" B 5, 1994, p. 264) [3, 11]; extended reaction time of 100 min or higher temperatures gave only $(\mu-H)_2Os_3(\mu_2-\eta^3-C_6H_4CH_2NC_5H_4N)(CO)_9$ [11]
27 $(\mu-H)Os_3(CO)_{10}(\mu_2-\eta^2-NC_3H_3N)$ Formula IV	from $Os_3(CO)_{10}(NCCH_3)_2$ (see "Organoosmium Compounds" B 3, 1994, p. 218) and a slight excess of pyrazole in refluxing benzene for 1 h; workup by TLC with CH_2Cl_2/petroleum ether as eluant (48%), together with 22% of $(\mu-H)Os_3(\mu_2-\eta^2-C_3H_3N_2)(CO)_{10}$ as the minor isomer (see "Organoosmium Compounds" B 5, 1994, p. 216) [13] yellow crystals (from pentane/CH_2Cl_2) [13] ^1H NMR (CDCl$_3$): -13.47 (s, OsH), 5.91 (d, H-4), 7.16 (d, H-2,5); $^3J=3.3$ [13] IR (cyclohexane): 1964 w, 1983 m, 1992 m, 2007 m, 2018 s, 2023 vs, 2061 s, 2070 vs, 2110 m (all CO) [13] EI mass spectrum: $[M]^+$, $[M-n\ CO]^+$, $n=1$ to 10 [13]
28 $Os_3(CO)_{10}(\mu_2-\eta^2-NC_4H_4N)$ compare Formula IV	from $Os_3(CO)_{12}$ and an excess of pyridazine in THF for 4 h in the presence of 2 equivalents of $(CH_3)_3NO\cdot2\ H_2O$; workup by filtration and removal of the solvent [1] ^{13}C NMR (THF/acetone-d$_6$, -89 °C; for numbering, see Formula IV): 178.7 ($CO_{d,d'}$), 180.1 ($CO_{c,c'}$), 184.6 ($CO_{b,b'}$), 186.3 ($CO_{a,a'}$), 185.9, 191.1 (each 1 CO, $CO_{e,f}$; $J(C,C)=33$); variable-temperature ^{13}C NMR determinations between -89 and $+54$ °C indicated two separate but simultaneous exchange processes, an in-plane cycling of the six equatorial CO's via a triply-bridged intermediate, and a localized, axial-equatorial exchange on the unique Os center; a low-energy process with a coalescence temperature of about -30 °C probably involves

References on p. 90

Table 5 (continued)

No. compound	method of preparation (yield) properties and remarks

	CO-b,b′, c,c′, d,d′ and CO-e,f, whereas above +40 °C all CO groups scrambled to produce a one-line spectrum; spectra at various temperatures depicted [1] IR (CH_2Cl_2): 1978, 2007, 2020 sh, 2030, 2055, 2082 (all CO) [1]
*29 (μ-H)Os$_3$(CO)$_{10}$\{μ_2-η^2-N(=NC$_6$H$_4$)N\} Formula IV	from Os$_3$(CO)$_{10}$(NCCH$_3$)$_2$ (see "Organoosmium Compounds" B 3, 1994, p. 218) and benzo-triazole in CH_2Cl_2 at room temperature for 2 h; workup by TLC with ether/petroleum ether as eluant (90%), together with 5% of isomeric (μ-H)Os$_3$(μ_2-η^2-C$_6$H$_4$N$_3$)(CO)$_{10}$ (see "Organoosmium Compounds" B 5, 1994, p. 228) [22] yellow crystals (from hexane/CH_2Cl_2) [22] ^{1}H NMR (CDCl$_3$): −13.25 (s, OsH), 7.25 to 7.90 (m, C$_6$H$_4$) [22] ^{13}C NMR (CDCl$_3$): 114.4, 117.9, 125.0, 126.8 (dd's, each 1 C of C$_6$H$_4$; ^{3}J(C,H)=7.7, ^{1}J(C,H)=162 and 165), 171.4 (d, 1 CO, CO$_b$ or CO$_{b'}$; J=12.1), 171.7, 171.9 (dd's, each 1 CO, CO$_{c,c'}$; J=3.3 and 2.2), 172.9 (d, 1 CO, CO$_b$ or CO$_{b'}$; J=12.1), 173.7, 174.1 (s's, each 1 CO, CO$_{a,a'}$), 176.7, 176.9 (dd's, each 1 CO, CO$_{d,d'}$; J=2.2 and 3.3), 180.1, 181.3 (s's, each 1 CO, CO$_{e,f}$); spectrum depicted [22] ^{15}N NMR (no medium or reference given): 187.1, 244.4, 346.7 [22] IR (cyclohexane): 1988 m, 1995 m, 2008 s, 2021 vs, 2029 vs, 2064 s, 2076 vs, 2113 m (all CO) [22] EI mass spectrum: [M]$^+$, [M−n CO]$^+$, n=1 to 10 [22]
*30 (μ-H)Os$_3$(CO)$_{10}$(μ_2-η^2-N=NC$_6$H$_5$) Formula V	IV (85%) [19] from Os$_3$(CO)$_{10}$(NCCH$_3$)$_2$ and C$_6$H$_5$NHNH$_2$ in CH_2Cl_2 at 25 °C for 1 h; workup by TLC with petroleum ether as eluant (45%) [19] brown crystals (from CH_2Cl_2/pentane); air-stable as solid and in solution; easily soluble in common organic solvents [19] ^{1}H NMR (CDCl$_3$): −13.48 (s, OsH), 7.5 (br, C$_6$H$_5$) [19] ^{13}C \{^{1}H\} NMR (CDCl$_3$): 173.6 (2 C), 174.0, 175.1 (each 1 C), 175.6 (2 C), 176.9, 178.1, 181.5, 183.0 (each 1 C) [19]

References on p. 90

Table 5 (continued)

No. compound	method of preparation (yield) properties and remarks

*30 (continued)

IR (cyclohexane): 1990 m, 1998 m, 2006 s, 2022 s, 2030 s, 2063 sh, 2068 vs, 2112 m (all CO) [19]

EI mass spectrum: $[M]^+$, $[M-n\ CO]^+$, $n=1$ to 10 [19]

*31 $(\mu-H)Os_3(CO)_{10}(\mu_2-\eta^2-N=NC_6H_4F-4)$
Formula V

IV (85%) [19]

brown crystals (from CH_2Cl_2/pentane); air-stable as solid and in solution; easily soluble in common organic solvents [19]

^1H NMR ($CDCl_3$): -13.42 (s, OsH), 6.59 to 7.33 (AB portion of ABX, C_6H_4; $^3J(H,H)=9.3$, $^3J(H,F)=8.6$, $^4J(H,F)=3.6$) [19]

^{19}F NMR (vs. internal C_6H_5F): 2.92; the shielding parameter indicates that the $\mu_2-\eta^2-$bonded N=N group is a weak resonance donor to the fluorobenzene ring [19]

IR (cyclohexane): 1989 w, 1997 w, 2007 m, 2014 s, 2027 s, 2058 s, 2066 vs, 2109 m (all CO) [19]

EI mass spectrum: $[M]^+$, $[M-n\ CO]^+$, $n=1$ to 10 [19]

*32 $(\mu-H)Os_3(CO)_{10}(\mu_2-\eta^2-N=NC_6H_4CH_3-4)$
Formula V

IV (85%) [19]

brown crystals (from pentane); air-stable as solid and in solution; easily soluble in common organic solvents [19]

^1H NMR ($CDCl_3$): -13.3 (s, OsH), 2.46 (s, CH_3), 6.94 (d, H-3,5), 7.46 (d, H-2,6; $^3J=9.6$) [19]

^{15}N NMR ($CDCl_3$?; vs. $^{15}NH_3$): 693.5 [19]

IR (cyclohexane): 1480 (νN=N), 1987 m, 1995 m, 2003 s, 2020 s, 2027 s, 2061 sh, 2066 vs, 2109 m (all CO); the ν(N=N) stretching frequency depended more strongly on the coordination mode of the N=NR ligand than on the electronic character of the aryl group R and the substituents [19]

UV (n-heptane): depicted, but no values given; [19]

EI mass spectrum: $[M]^+$, $[M-n\ CO]^+$, $n=1$ to 10 [19]

33 $[(\mu-H)Os_3(CO)_{10}(\mu_2-\eta^2-N=NHC_6H_4CH_3-4)]BF_4$

by protonation of No. 32 with a slight excess of $HBF_4 \cdot O(C_2H_5)_2$ in CH_2Cl_2 and precipitation with pentane [19]

References on p. 90

Table 5 (continued)

No. compound	method of preparation (yield) properties and remarks
	^1H NMR (CD_2Cl_2): 15.4 (d, NH; $J(H,^{15}N)=81$) [19] ^{15}N NMR $(CD_2Cl_2?; vs. ^{15}NH_3)$: 319.5 (d); $J(H,^{15}N)=81$ [19]
*34 $(\mu-H)Os_3(CO)_{10}(\mu_2-\eta^2-NHN=NH)$ Formula VI	II (55%) [21] from $(\mu-H)_2Os_3(CO)_{10}$ and $(CH_3)_3SiN_3$ under the conditions of Preparation Method II (25%) [12] orange crystals (from hexane) [12] ^1H NMR (C_6D_6): -13.35 (s, OsH), 8.38 (s, NH) [21]; see also [12] IR (hexane): 1982 w, 1996 m, 2001 m, 2015 s, 2025 s, 2048 w, 2061 s, 2068 s, 2111 w (all CO), 3363 vw (NH) [21]; see also [12] EI mass spectrum: $[M]^+$ [12], $[M]^+$, $[M-n\,CO]^+$, n=1 to 9 [21] thermolysis in refluxing octene for 45 min [12] or hexane for 24 h [21] gave $(\mu-H)Os_3(CO)_{10}$-$(\mu_2-\eta^1-NH_2)$ (Section 3.1.1.7.2.1) [12, 21]
35 $(\mu-H)Os_3(CO)_{10}(\mu_2-\eta^2-NHN=NC_4H_9-n)$ Formula VI	II (67%) [8, 9] yellow crystals [8] from aqueous ethanol [9] ^1H NMR $(CDCl_3)$: -12.50 (s, OsH), 1.04 to 2.01 (m, CH_3?), 3.96 to 4.31 (m, CH_2?), 8.01 (br s, NH) [9] IR (hexane): 1978 w, 1990 m, 1998 m, 2007 m, 2014 s, 2025 s, 2031 m, 2045 w, 2050 s, 2067 s, 2109 w (all CO) [9] mass spectrum: $[M]^+$ [9] thermolysis in refluxing 1,2-dimethoxyethane resulted in $(\mu-H)_2Os_3(CO)_9(\mu_3-\eta^1-NC_4H_9-n)$ (Section 3.1.1.7.2.4) [8, 9]
36 $(\mu-H)Os_3(CO)_{10}(\mu_2-\eta^2-NHN=NC_6H_{11}-cyclo)$ Formula VI	II (43%) [8, 9] yellow crystals (from hexane) [8, 9] ^1H NMR $(CDCl_3)$: -12.52 (s, OsH), 1.45 to 2.02 (m, CH_2), 3.96 (m, CH of C_6H_{11}-cyclo), 8.02 (br s, NH) [9] IR (cyclohexane): 1976 w, 1988 w, 1997 w, 2006 w, 2015 s, 2024 s, 2045 w, 2059 s, 2065 s, 2109 w (all CO) [9] EI mass spectrum: $[M]^+$ [9] thermolysis in refluxing 1,2-dimethoxyethane gave $(\mu-H)_2Os_3(CO)_9(\mu_3-\eta^1-NC_6H_{11}-cyclo)$ (Section 3.1.1.7.2.4) [8, 9]

References on p. 90

Table 5 (continued)

No. compound	method of preparation (yield) properties and remarks

37 $(\mu-H)Os_3(CO)_{10}(\mu_2-\eta^2-NHN=NCH_2C_6H_5)$
  Formula VI

II (65%) [8, 9]
yellow crystals (from hexane) [8, 9]
^1H NMR (CD_2Cl_2): -12.77 (s, OsH), 4.65, 5.36
  (d's, each 1 H of CH_2; $^2J=13$), 8.13 (br s,
  NH) [9]
IR (hexane): 1978 w, 1991 w, 2001 m, 2010 s,
  2014 s, 2025 m, 2045 w, 2061 s, 2067 s,
  2108 w (all CO) [9]
EI mass spectrum: $[M]^+$ [9]
thermolysis in refluxing 1,2-dimethoxyethane
  gave $(\mu-H)_2Os_3(CO)_9(\mu_3-\eta^1-NCH_2C_6H_5)$ (Section 3.1.1.7.2.4) [8, 9]

38 $(\mu-H)Os_3(CO)_{10}\{\mu_2-\eta^2- NHN=NC(=CH_2)C_6H_5\}$
  Formula VI

II (61%) [8, 9]
yellow crystals (from hexane) [8, 9]
^1H NMR (acetone-d_6): -12.62 (s, OsH), 4.80,
  5.05 (s's, each 1 H of CH_2), 10.40 (br s, NH);
  no resonances for C_6H_5 given [9]
IR (hexane): 1978 w, 1989 w, 2001 m, 2008 sh,
  2012 s, 2025 s, 2044 w, 2058 s, 2068 s,
  2109 w (all CO) [9]
EI mass spectrum: $[M]^+$ [9]
thermolysis in refluxing 1,2-dimethoxyethane
  for 24 h yielded $(\mu-H)Os_3(CO)_{10}\{\mu_2-\eta^1-N=C(CH_3)C_6H_5\}$ (Section 3.1.1.7.2.3); a
  possible reaction mechanism was discussed [9]

*39 $(\mu-H)Os_3(CO)_{10}(\mu_2-\eta^2-NHN=NC_6H_5)$
  Formula VI

II (81%) [8, 9]
yellow crystals (from hexane) [8, 9]
^1H NMR (CD_2Cl_2): -12.41 (s, OsH), 7.04 to
  7.55 (m, C_6H_5), 8.75 (br s, NH) [9]; similar
  data in $CDCl_3$ [8, 21]
IR (hexane): 1975 w, 1990 w, 2001 m, 2010 sh,
  2015 s, 2026 s, 2047 w, 2061 s, 2069 s,
  2110 w (all CO) [8, 9, 21]
EI mass spectrum: $[M]^+$ [9]
thermolysis in refluxing 1,2-dimethoxyethane
  led to $(\mu-H)_2Os_3(CO)_9(\mu_3-\eta^1-NC_6H_5)$ (Section
  3.1.1.7.2.4) [8, 9]

References on p. 90

Table 5 (continued)

No. compound	method of preparation (yield) properties and remarks
*40 $Os_3(CO)_{12}\{\mu_2\text{-}\eta^2\text{-}(NCO_2CH_3)_2\}$	by carbonylation of

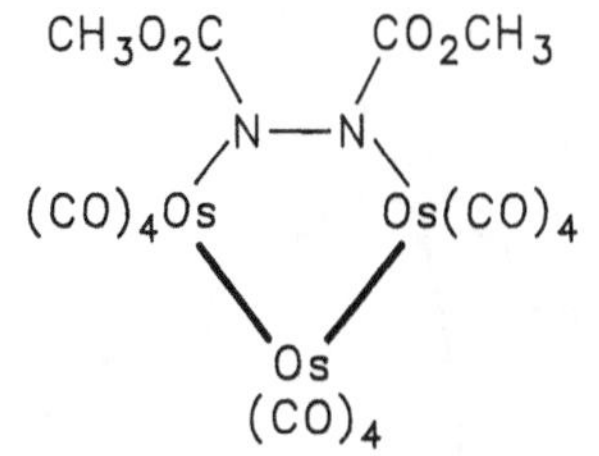

$$CH_3O_2C \qquad CO_2CH_3$$
$$\diagdown \qquad \diagup$$
$$N\text{---}N$$
$$\diagup \qquad \diagdown$$
$$(CO)_4Os \qquad Os(CO)_4$$
$$\diagdown \qquad \diagup$$
$$Os$$
$$(CO)_4$$

by carbonylation of
$Os_3(CO)_{10}\{\mu_2\text{-}\eta^4\text{-}(NCO_2CH_3)_2\}$ (Section
3.1.1.7.4) in CH_2Cl_2 at room temperature for
12 to 24 h under 102 atm of CO (quantitative
yield, based on IR and mass spectra) [16,
17]

by carbonylation of $Os_3(CO)_{11}\{\mu_2\text{-}\eta^3\text{-}$
$(NCO_2CH_3)_2\}$ (Section 3.1.1.7.4) in CH_2Cl_2 at
0 °C for 30 h under 1 atm of CO (quantitative
yield, based on IR spectrum) [16, 17]

yellow, air-stable crystals (from
hexane/CH_2Cl_2); m.p. 126 °C [16, 17]
^1H NMR ($CDCl_3$): 3.64 (s, CH_3) [16, 17]
IR (C_2Cl_4): 1443 mw, 1633 m (CO_2CH_3);
(hexane): 1995 mw, 2026 m, 2062 s, 2075 s,
2123 s (all CO) [16, 17]
FAB mass spectrum (3-mercapto-1,2-propane-
diol): $[M]^+$ [16, 17]

*Further information:

$Os_3(CO)_{11}\{\mu_2\text{-}\eta^2\text{-}N(C_6F_5)O\}$ (Table 5, No. 1) crystallizes in the triclinic space group
$P\bar{1} - C_i^1$ (No. 2) with a = 7.7080(10), b = 9.482(2), c = 17.079(3) Å, α = 81.760(1)°, β = 79.520(2)°,
γ = 74.060(4)°; Z = 2, D_c = 3.042 g/cm³. The structure is shown in **Fig. 25**. The cluster consists
of an open Os_3 triangle with an elongated nonbonding Os(1)–Os(2) distance of 3.59 Å, which
is bridged by the nitrogen atom of the $\mu_2\text{-}\eta^2\text{-}N(C_6F_5)O$ ligand. In addition, the crystal exhibits
a unique triangle formed by the Os(2) center and the N and O atoms of the $\mu_2\text{-}\eta^2\text{-}N(C_6F_5)O$
ligand acting as a four-electron donor. The N–O bond of 1.944 Å is indicative of predominant
single-bond character. The Os(3)–Os(1)–N and Os(3)–Os(2)–N bond angles amount to 81.5(3)°
and 83.0(4)°, respectively, the Os(3)–Os(2)–O angle is 86.2(4)° [25].

$(\mu\text{-H})Os_3(CO)_{10}\{\mu_2\text{-}\eta^2\text{-}N(C_6H_4CH_3\text{-}4)CH\text{=}O\}$ (Table 5, No. 9) crystallizes in the triclinic
space group $P\bar{1} - C_i^1$ (No. 2) with a = 7.295(2), b = 11.705(3), c = 13.928(4) Å, α = 66.71(2)°,
β = 80.17(2)°, γ = 79.18(2)°; Z = 2, D_c = 2.83 g/cm³. The structure is shown in **Fig. 26**. Both
the $\mu\text{-H}$ ligand and the hydrogen atom H(1) of the $\mu_2\text{-}\eta^2\text{-}N(C_6H_4CH_3\text{-}4)CH\text{=}O$ ligand were
not observed crystallographically; H(1) is shown in an idealized position and the hydride
ligand is believed to span the Os(1)–Os(3) bond edge. The formamido ligand is bonded
in a diaxial coordination arrangement; the ligand plane is nearly perpendicular to the Os_3
triangle; the dihedral angle is 74°. The distances in the N–C–O moiety are intermediate
between single and double bonds, indicating electron delocalization across the ligand [2, 4].

$(\mu\text{-H})Os_3(CO)_9\{\mu_2\text{-}\eta^2\text{-}NH_2CH(CO_2C_2H_5)CH_2O\}$ (Table 5, No. 12) crystallizes in the triclinic
space group $P\bar{1} - C_i^1$ (No. 2) with a = 8.898(3), b = 12.832(4), c = 10.022(2) Å, α = 96.77°, β =
103.74°, γ = 90.08°; Z = 2, D_c = 2.877 g/cm³. The structure is shown in **Fig. 27**. The compound
is isostructural with No. 19 having a tridentate chelating ligand derived from serine and
being asymmetrically coordinated relative to the metal unit in a $\mu_2\text{-}\eta^2$ mode. The symmetri-

References on p. 90

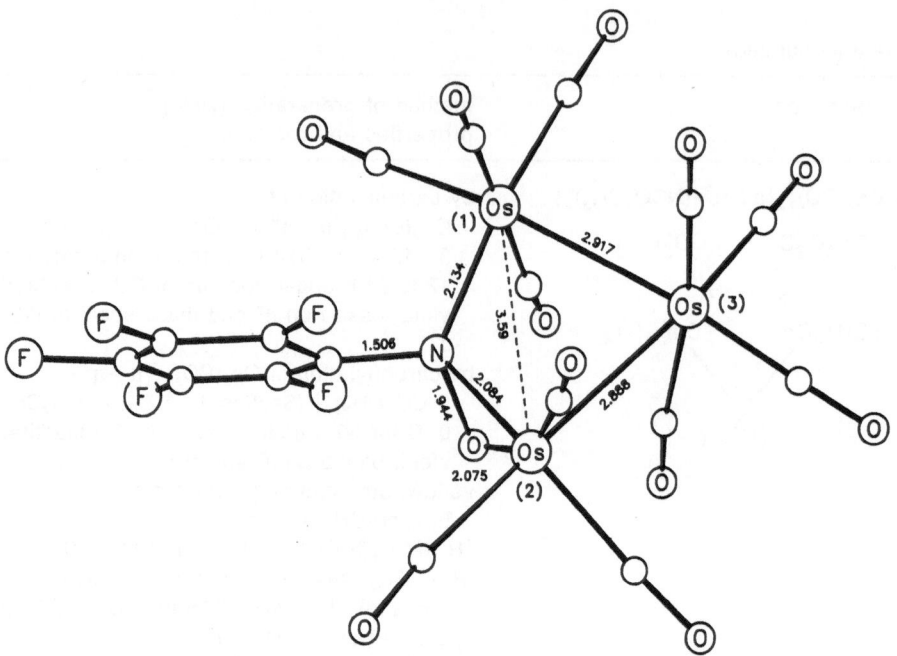

Fig. 25. Molecular structure of $Os_3(CO)_{11}\{\mu_2-\eta^2-N(C_6F_5)O\}$ (No. 1) with selected interatomic distances (in Å) [25].

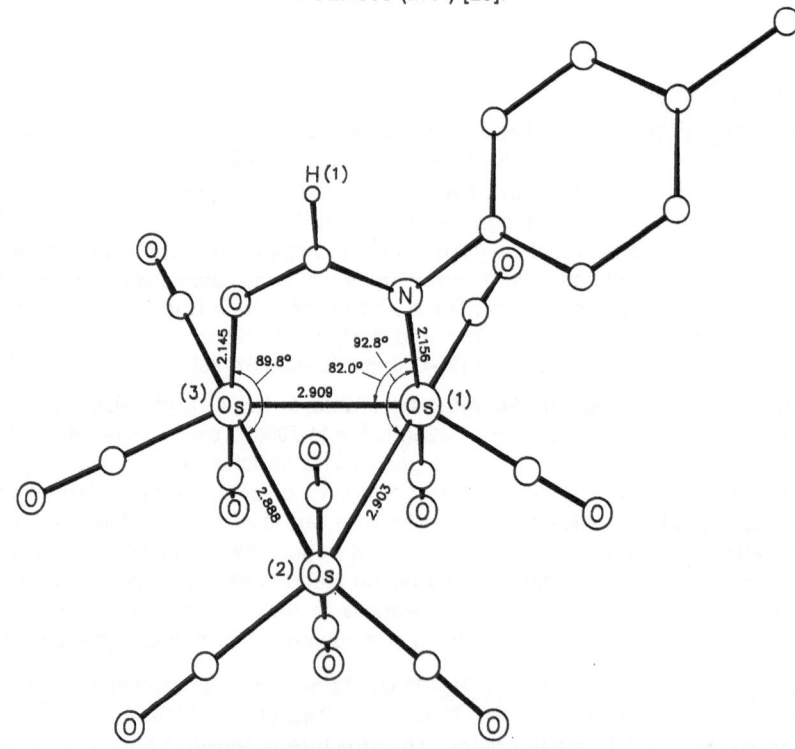

Fig. 26. Molecular structure of $(\mu-H)Os_3(CO)_{10}\{\mu_2-\eta^2-N(C_6H_4CH_3-4)CH=O\}$ (No. 9) with selected bond distances (in Å) and bond angles [4].

References on p. 90

cally bridging oxygen atom deviates from the Os₃ plane by 1.56 Å; the coordination environment is trigonal-pyramidal. The five-membered chelate ring formed by the Os(1)-NH₂-CH-CH₂-O atoms exhibits envelope conformation with C(2) deviating by 0.65 Å from the O-C(1)-N-Os(1) plane. The CO₂C₂H₅ group exhibits pseudoaxial orientation relative to the Os₃ framework; the angle between the Os₃ plane and the C(3)-O(1)-O(2) plane is 84.3°. The orientation of the planes of the carboxyl and ethoxy group is perpendicular. The μ-H ligand and hydrogen atoms were not observed crystallographically. Average bond lengths of Os-CO are 1.90 Å, while the C-O distances range from 1.11 to 1.19 Å; similar values were also observed for No. 13, see below [24].

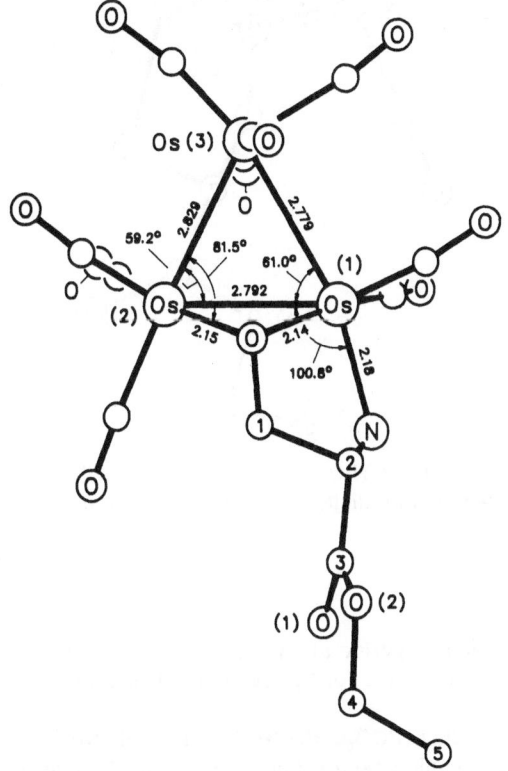

Fig. 27. Molecular structure of (μ-H)Os₃(CO)₉{μ₂-η²-NH₂CH(CO₂C₂H₅)CH₂O} (No. 12) with selected bond distances (in Å) and bond angles [24].

(μ-H)Os₃(CO)₉{μ₂-η²-NH₂CH(CO₂C₂H₅)CH₂O} (Table 5, No. 13) crystallizes in the monoclinic space group P2₁/n−C⁵₂ₕ (No. 14) with a=9.371(1), b=23.346(5), c=9.798(2) Å, β= 91.10(2)°; Z=4, D_c=2.962 g/cm³. The structure, shown in **Fig. 28**, differs from that of No. 12 mainly by the conformation of the five-membered chelate ring which is twisted with deviation of C(1) and C(2) by −0.32 and +0.26 Å, respectively, from the plane traversing μ-O, Os(1), and N. Furthermore, the ester group is equatorially oriented relative to Os₃; the angle between the Os₃ plane and the C(3)-O(1)-O(2) plane is 35.3°. Contrary to Nos. 12 and 19, the orientation of the planes of the carboxyl and ethoxy groups is parallel. The hydrogen

References on p. 90

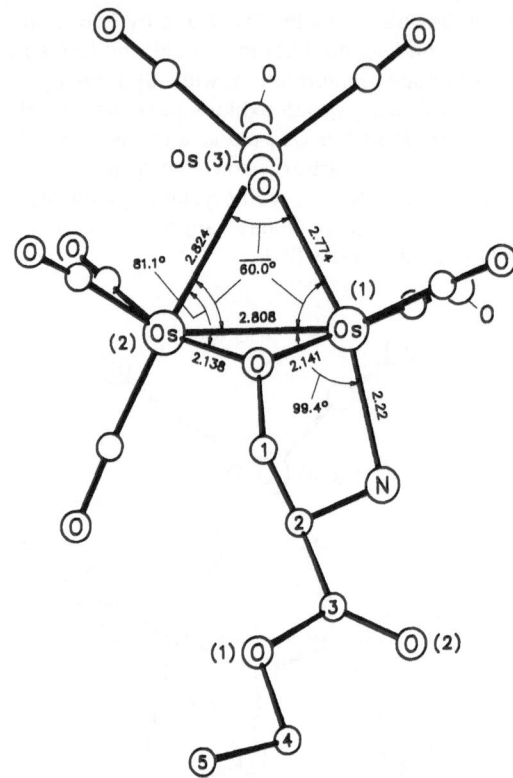

Fig. 28. Molecular structure of (μ-H)Os$_3$(CO)$_9${μ_2-η^2-NH$_2$CH(CO$_2$C$_2$H$_5$)CH$_2$O} (No. 13) with
selected bond distances (in Å) and bond angles [24].

atoms, including the bridging hydride atom at the Os(1)-Os(2) bond, were located crystallographically but not given in Fig. 28. For Os-CO and C-O distances, see No. 12 [24].

(μ-H)Os$_3$(CO)$_9${μ_2-η^2-NH$_2$CH(CO$_2$C$_2$H$_5$)CH$_2$S} (Table 5, No. 19) crystallizes in form of
yellow platelets in the triclinic space group P$\bar{1}$−C$_i^1$ (No. 2) with a=8.837(2), b=10.225(5),
c=13.068(5) Å, α=96.99(3)°, β=90.88(2)°, γ=106.03(3)°; Z=2, D$_c$=2.871 g/cm^3. The structure is shown in **Fig. 29**. The tridentate chelating μ_2-η^2-NH$_2$CH(CO$_2$C$_2$H$_5$)CH$_2$S ligand corresponds to an envelope conformation, formed by a single plane of the S, N, C(1), and Os(1)
atoms, while C(2) deviates from this plane by 0.68 Å. The symmetrically bridging sulfur
atom spanning the Os(1)-Os(2) bond edge is disposed by 1.90 Å out of the metallacycle
plane; the coordination environment around the S atom is trigonal-pyramidal. The NH$_2$
group exhibits strong coordination to the Os(1) center; the considerable reduction of the
Os(1)-Os(3) bond is attributed to a trans effect of the amino group. The C(3)-O(1)-O(2)
plane is almost perpendicularly oriented to the Os$_3$ plane; the angle between these planes
amounts to 73.2°. The carbon atoms of the C$_2$H$_5$ moiety are randomly disordered and occupy
the two crystallographically independent positions C(4)-C(5) and C(4)'-C(5)'. The Os-CO
bond lengths are 1.90 Å on average, the C-O distances range from 1.07 to 1.23 Å. The
μ-H ligand and hydrogen atoms were not observed crystallographically [24].

References on p. 90

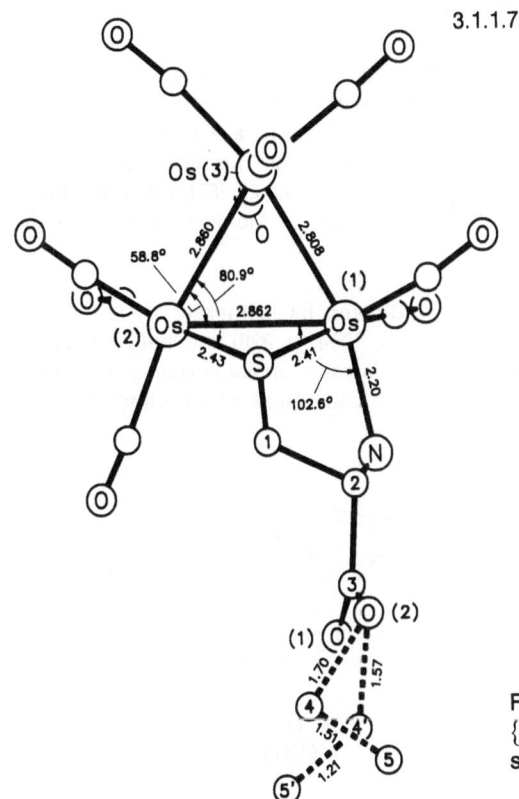

Fig. 29. Molecular structure of $(\mu\text{-H})Os_3(CO)_9$- $\{\mu_2\text{-}\eta^2\text{-}NH_2CH(CO_2C_2H_5)CH_2S\}$ (No. 19) with selected bond distances (in Å) and bond angles [24].

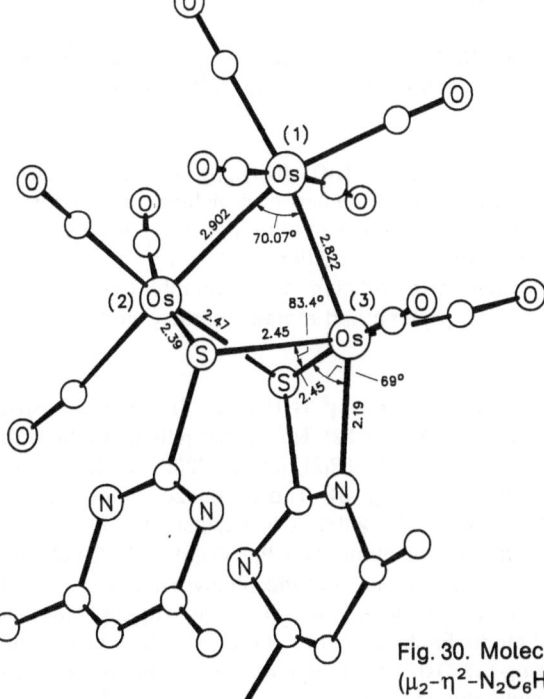

Fig. 30. Molecular structure of Molecule 1 of $Os_3(CO)_9$- $(\mu_2\text{-}\eta^2\text{-}N_2C_6H_7S)(\mu\text{-}SC_6H_7N_2)$ (No. 20) with selected bond distances (in Å) and bond angles [27].

References on p. 90

Os$_3$(CO)$_9$(μ_2-η^2-N$_2$C$_6$H$_7$S)(μ-SC$_6$H$_7$N$_2$) (Table **5**, No. **20**) crystallizes in the monoclinic space group P2$_1$/n – C$_{2h}^5$ (No. 14) with a = 13.918(4), b = 23.718(8), c = 18.631(7) Å, β = 103.93(3)°; Z = 4, D$_c$ = 2.45 g/cm^3. Within the asymmetric unit there are two independent molecules of No. 20; the structure of Molecule 1 is shown in **Fig. 30**. The molecule consists of an open Os$_3$ framework; the non-bonding Os(2)···Os(3) amounts to 3.287(3) Å. The interatomic S···S bond amounts 3.26(2) Å. Most of the bond lengths in Molecule 2 are slightly longer than in Molecule 1 [27].

(μ-H)Os$_3$(CO)$_{10}$\{μ_2-η^2-N(=NC$_6$H$_4$)N\} (Table **5**, No. **29**) and its isomer (μ-H)Os$_3$(μ_2-η^2-C$_6$H$_4$N$_3$)(CO)$_{10}$ (see "Organoosmium Compounds" B 5, 1994, p. 228) are believed to be formed in corrosion inhibition processes by treatment of corroded metal surfaces with benztriazole. Two isomeric forms were discussed for the title compound, see Formula VII; the ^{15}N NMR data unambiguously favor structure VIIa [22].

VII

Cyclic voltammetry in CH$_2$Cl$_2$ (c = 5 × 10^{-4} M, 0.1 M [N(C$_4$H$_9$-n)$_4$]BF$_4$, Pt electrode, – 10 °C) showed a chemically irreversible reduction at – 1.52 V vs. SCE, and a small oxidation peak at – 0.36 V vs. SCE [22].

Protonation with CF$_3$CO$_2$H in CDCl$_3$ led to **[(μ-H)Os$_3$(CO)$_{10}$\{μ_2-η^2-N(=NC$_6$H$_4$)NH\}]-[CF$_3$CO$_2$]**, based on ^{1}H NMR and IR spectroscopy, exhibiting a stretching N-H absorption at 3060 cm^{-1}, while no changes in the μ-H region and only minor differences in the aromatic region were observed in the ^{1}H NMR spectrum. The addition of the hydrogen to one of the nitrogens bearing a lone pair is indicative for an efficient corrosion inhibition by benzotriazole since it prevents the contact between protons and metallic surface and the subsequent reaction [22].

(μ-H)Os$_3$(CO)$_{10}$(μ_2-η^2-N=NC$_6$H$_5$), **(μ-H)Os$_3$(CO)$_{10}$(μ_2-η^2-N=NC$_6$H$_4$F-4)**, and **(μ-H)Os$_3$-(CO)$_{10}$(μ_2-η^2-N=NC$_6$H$_4$CH$_3$-4)** (Table **5**, Nos. **30**, **31**, and **32**). No. 30 crystallizes in the orthorhombic space group P2$_1$2$_1$2$_1$ – D$_2^4$ (No. 19) with a = 9.482(2), b = 12.542(3), c = 17.588(4) Å; Z = 4, D$_c$ = 3.04 g/cm^3. The structure is shown in **Fig. 31**. The aryldiazo ligand occupies a diaxial μ_2-η^2-bidging site; the N-N distance of 1.20(4) Å is consistent with a double bond. The dihedral angle between the triosmium plane and the Os(2)-N-N-Os(3) plane is 103.25°; the dihedral angle between the latter plane and the phenyl ring is 15.76°. The μ-H ligand was not located directly but it is believed to bridge the slightly elongated Os(2)-Os(3) bond edge in a diequatorial mode. Each of the three Os atoms show an essential octahedral stereochemistry upon ignoring the Os(2)-Os(3) linkage; the octahedra centered at Os(2) and Os(3) are rotated contrafacially due to the μ_2-η^2-N=NC$_6$H$_5$ ligand [19].

References on p. 90

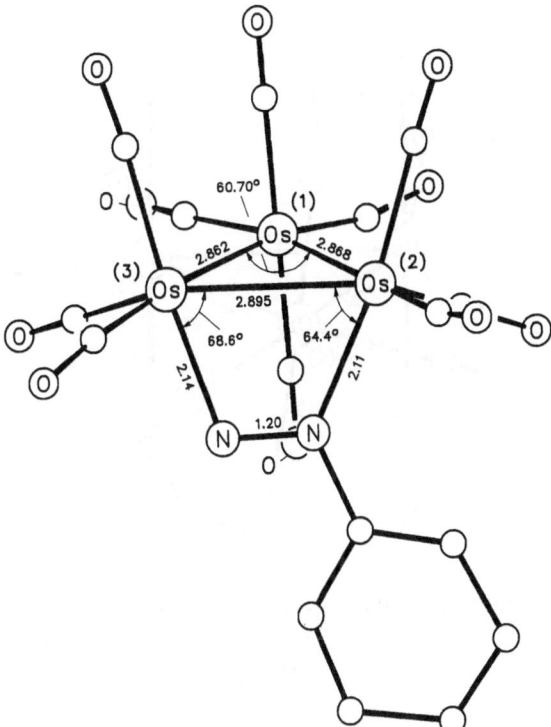

Fig. 31. Molecular structure of $(\mu\text{-}H)Os_3(CO)_{10}(\mu_2\text{-}\eta^2\text{-}N{=}NC_6H_5)$ (No. 30) with selected bond distances (in Å) and bond angles [19].

Both photolysis and thermolysis of Nos. 30, 31, and 32 led to $(\mu\text{-}H)Os_3(CO)_{10}(\mu_2\text{-}\eta^1\text{-}N{=}NC_6H_4R\text{-}4)$; R=H, F, CH_3 (Section 3.1.1.7.2.3). UV irradiation under the conditions of Preparation Method IV (450-W medium-pressure mercury-vapor lamp; see p. 66) gave yields of ca. 15%, while refluxing in heptane in the absence of CO led to yields of more than 95%, based on ^1H NMR spectra. The conversions proceeded likewise under CO, H_2, N_2, or Ar and were also unaffected by the presence of $P(C_6H_5)_3$. Extended irradiation resulted in decomposition [19].

Protonation with gaseous HCl or $HBF_4 \cdot O(C_2H_5)_2$ in CH_2Cl_2 gave an deep-blue solution which was stable in the case of $HBF_4 \cdot O(C_2H_5)_2$ but slowly reversed in the case of HCl; addition of pentane precipitated the corresponding salts probably of the type $[(\mu\text{-}H)Os_3\text{-}(CO)_{10}(\mu_2\text{-}\eta^2\text{-}N{=}NHC_6H_4R\text{-}4)]X$ (for R=H, F, and CH_3; X=Cl or BF_4) which could not be isolated in pure form. Similar protonation of No. 32 (R=CH_3) with $HBF_4 \cdot O(C_2H_5)_2$ gave No. 33 [19].

$(\mu\text{-}H)Os_3(CO)_{10}(\mu_2\text{-}\eta^2\text{-}NHN{=}NH)$ (Table 5, No. 34) crystallizes in the triclinic space group $P\bar{1}-C_i^1$ (No. 2) with a=7.867(3), b=9.113(3), c=14.000(5) Å, α=90.68(2)°, β=100.37(2)°, γ= 118.80(2)°; Z=2, D_c=3.46 g/cm³. The structure is shown in **Fig. 32**. The structure can be regarded as derived from $Os_3(CO)_{12}$ by replacing two adjacent cis-axial carbonyls by the symmetrically chelating triazenido ligand. Both the μ-H ligand and NH hydrogens on N(1) and N(2) were not located directly, but the hydride is believed to span the elongated Os(2)-Os(3) bond edge and to lie in the Os_3 plane. The dihedral angle between the Os(2)-N(1)-N(3)-N(2)-Os(3) plane and the Os_3 plane amounts to 83.5° [12].

                     References on p. 90

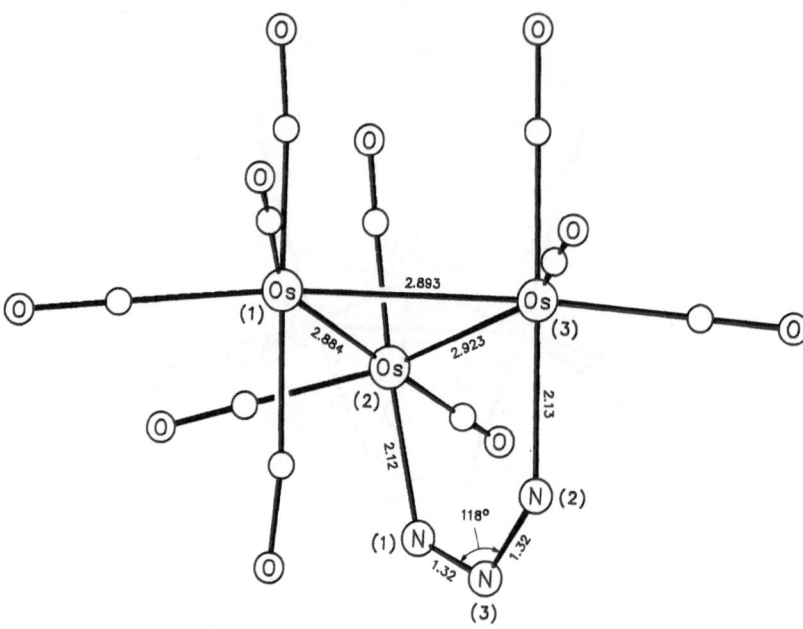

Fig. 32. Molecular structure of $(\mu-H)Os_3(CO)_{10}(\mu_2-\eta^2-NHN=NH)$ (No. 34) with selected bond distances (in Å) and a bond angle [12].

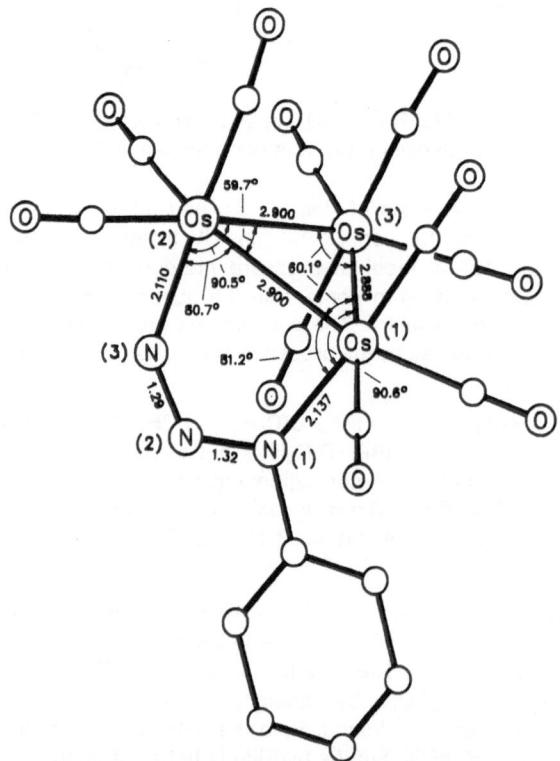

Fig. 33. Molecular structure of $(\mu-H)Os_3(CO)_{10}(\mu_2-\eta^2-NHN=NC_6H_5)$ (No. 39) with selected bond distances (in Å) and bond angles [8, 9].

References on p. 90

(μ-H)Os$_3$(CO)$_{10}$(μ$_2$-η^2-NHN=NC$_6$H$_5$) (Table **5**, No. **39**) crystallizes in the monoclinic space group P2$_1$/c − C$_{2h}^5$ (No. 14) with a = 9.665(3), b = 8.823(4), c = 25.571(14) Å, β = 100.17(4)°; Z = 4, D$_c$ = 3.01 g/cm. The structure is shown in **Fig. 33**. The Os atoms define an equilateral triangle. The Os(1)-Os(2) bond edge is bridged by the triazenido ligand with the N(1) and N(3) atoms occupying axial coordination sites on the same side of the Os$_3$ plane. The two N-N distances are equivalent, indicating electron delocalization over the symmetrically bridging μ$_2$-η^2- NHN=NC$_6$H$_5$ unit. The hydride was not located directly but is believed to bridge the Os(1)- Os(2) edge lying in the plane of the Os$_3$ triangle as concluded from the distribution of the carbonyl groups on the bridgehead metal centers. The Os(1)-N(1)-N(2)-N(3)-Os(2) unit is planar and makes an angle of 82.9(1)° with the Os$_3$ triangle, whereas the dihedral angle between the first plane and the phenyl ring amounts to 55.7° [8, 9].

Os$_3$(CO)$_{12}${μ$_2$-η^2-(NCO$_2$CH$_3$)$_2$} (Table **5**, No. **40**) crystallizes in the monoclinic space group P2$_1$/c − C$_{2h}^5$ (No. 14) with a = 12.136(2), b = 12.298(2), c = 16.027(3) Å, β = 91.37(1)°; Z = 4, D$_c$ = 2.924 g/cm^3. The structure, shown in **Fig. 34**, can be viewed as derived from Os$_3$(CO)$_{12}$ by the replacement of one Os-Os bond by the bridging two-electron donating μ$_2$-η^2- (NCO$_2$CH$_3$)$_2$ ligand; the N-N distance of 1.411(7) Å clearly indicates a single bond. The triangular disposition of the metal atoms is preserved with a nonbonding Os···Os distance of 4.198 Å [16, 17].

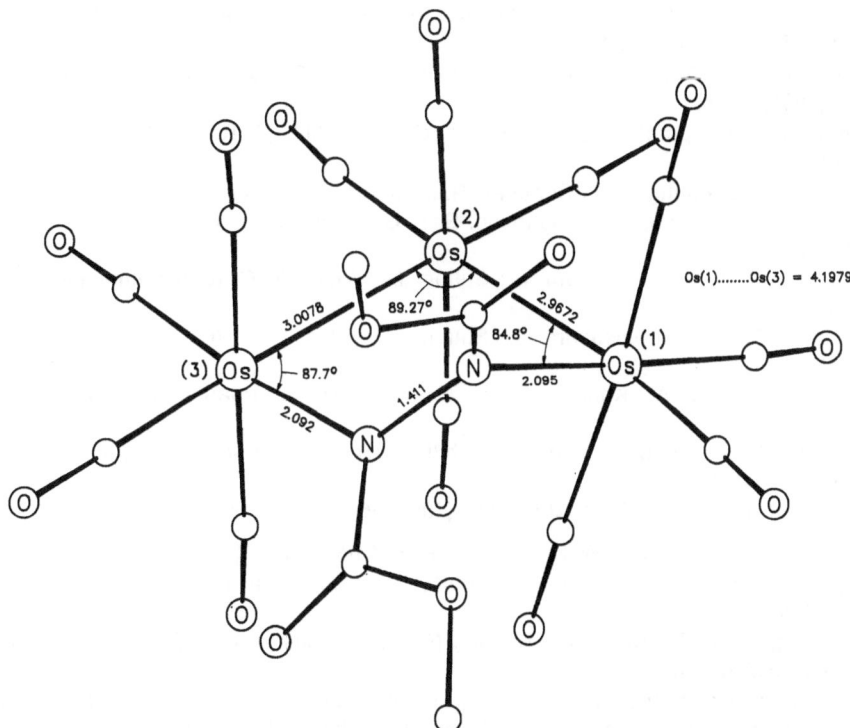

Os(1)........Os(3) = 4.1979

Fig. 34. Molecular structure of Os$_3$(CO)$_{12}${μ$_2$-η^2-(NCO$_2$CH$_3$)$_2$} (No. 40) with selected inter- atomic distances (in Å) and bond angles [17].

Decomposed under UV irradiation in CH$_2$Cl$_2$. Thermolysis in benzene at 70 °C for 5 h gave Os$_3$(CO)$_{10}${μ$_2$-η^4-(NCO$_2$CH$_3$)$_2$} (Section 3.1.1.7.4) [17].

                    References on p. 90

Treatment with an excess of $(CH_3)_3NO$ (dissolved in CH_3OH) in benzene at room temperature gave $Os_3(CO)_{10}\{\mu_2\text{-}\eta^4\text{-}(NCO_2CH_3)_2\}$ as the main product besides $Os_3(CO)_{11}\{\mu_2\text{-}\eta^3\text{-}(NCO_2CH_3)_2\}$ (both Section 3.1.1.7.4) [17].

References:

[1] Cotton, F. A.; Hanson, B. E. (Inorg. Chem. **16** [1977] 2820/2).
[2] Adams, R. D.; Golembeski, N. M. (J. Organomet. Chem. **171** [1979] C 21/C 26).
[3] Deeming, A. J.; Peters, R. (J. Organomet. Chem. **202** [1980] C 39/C 42).
[4] Adams, R. D.; Golembeski, N. M.; Selegue, J. P. (Inorg. Chem. **20** [1981] 1242/7).
[5] Johnson, B. F. G.; Lewis, J.; Odiaka, T. I.; Raithby, P. R. (J. Organomet. Chem. **216** [1981] C 56/C 60).
[6] Lin, Y. C.; Knobler, C. B.; Kaesz, H. D. (J. Am. Chem. Soc. **103** [1981] 1216/8).
[7] Burgess, K.; Johnson, B. F. G.; Lewis, J. (J. Organomet. Chem. **233** [1982] C 55/C 58).
[8] Burgess, K.; Johnson, B. F. G.; Lewis, J.; Raithby, P. R. (J. Organomet. Chem. **224** [1982] C 40/C 44).
[9] Burgess, K.; Johnson, B. F. G.; Lewis, J.; Raithby, P. R. (J. Chem. Soc. Dalton Trans. **1982** 2085/92).
[10] Deeming, A. J.; Peters, R. (J. Organomet. Chem. **235** [1982] 221/9).

[11] Deeming, A. J.; Peters, R.; Hursthouse, M. B.; Backer-Dirks, J. D. (J. Chem. Soc. Dalton Trans. **1982** 1205/11).
[12] Johnson, B. F. G.; Lewis, J.; Raithby, P. R.; Sankey, S. W. (J. Organomet. Chem. **228** [1982] 135/8).
[13] Shapley, J. R.; Samkoff, D. E.; Bueno, C.; Churchill, M. R. (Inorg. Chem. **21** [1982] 634/9).
[14] Arce, A. J.; Deeming, A. J.; Shaunak, R. (J. Chem. Soc. Dalton Trans. **1983** 1023/5).
[15] Burgess, K.; Holden, H. D.; Johnson, B. F. G.; Lewis, J. (J. Chem. Soc. Dalton Trans. **1983** 1199/202).
[16] Einstein, F. W. B.; Nussbaum, S.; Sutton, D.; Willis, A. C. (Organometallics **2** [1983] 1259/61).
[17] Einstein, F. W. B.; Nussbaum, S.; Sutton, D.; Willis, A. C. (Organometallics **3** [1984] 568/74).
[18] Lin, Y.-C.; Mayr, A.; Knobler, C. B.; Kaesz, H. D. (J. Organomet. Chem. **272** [1984] 207/29).
[19] Samkoff, D. E.; Shapley, J. R.; Churchill, M. R.; Wasserman, H. J. (Inorg. Chem. **23** [1984] 397/402).
[20] Odiaka, T. I. (J. Organomet. Chem. **284** [1985] 95/9).

[21] Smieja, J. A.; Gladfelter, W. L. (J. Organomet. Chem. **297** [1985] 349/59).
[22] Aime, S.; Botta, M.; Gobetto, R.; Osella, D.; Padovan, F. (J. Organomet. Chem. **353** [1988] 251/7).
[23] Deeming, A. J.; Meah, M. N.; Randle, N. P.; Hardcastle, K. I. (J. Chem. Soc. Dalton Trans. **1989** 2211/6).
[24] Podberezskaya, N. V.; Virovets, A. V.; Slovokhotov, Y. L.; Struchkov, Y. T.; Maksakov, V. A.; Kirin, V. P.; Ershova, V. A. (Zh. Strukt. Khim. **31** [1990] 108/16; J. Struct. Chem. [Engl. Transl.] **31** [1990] 943/50).
[25] Ang, H. G.; Kwik, W. L.; Ong, K. K. (J. Fluorine Chem. **60** [1993] 43/8).
[26] Maksakov, V. A.; Kirin, V. P.; Ershova, V. A.; Tkachev, S. V.; Semyannikov, P. P. (Izv. Akad. Nauk Ser. Khim. **1993** 1981/4; Russ. Chem. Bull. [Engl. Transl.] **42** [1993] 1898/901).
[27] Au, Y.-K.; Cheung, K.-K.; Wong, W.-T. (J. Chem. Soc. Dalton Trans. **1995** 1047/57).

3.1.1.7.3.3 Compounds with μ_3-η^2-Bonded N Ligands

Most of the compounds described in this section consist of a $(\mu$-H)Os$_3$(CO)$_9$ skeleton containing a face-capping μ_3-η^2 ligand bonded by N and O, N and S, or two N atoms. The compounds are represented by Formulas I and II. Some of the compounds deviating slightly from these general structures are represented by individual figures in Table 6. In Os$_3$(CO)$_9$(μ_3-η^2-SC$_5$H$_4$N)(μ-SC$_5$H$_4$N) (No. 3) the μ-H ligand is replaced by a bridging 2-pyridyl sulfide ligand.

The compounds of the type $(\mu$-H)Os$_3$(CO)$_9$(μ_3-η^2-SC$_3$H$_4$NE$'$) and $(\mu$-H)Os$_3$(CO)$_9$(μ_3-η^2-SC$_7$H$_4$NE$'$) (E$'$=O, S, or NH; Nos. 12, 13, 14, and 15) were obtained as inseparable mixtures of two isomers differing in the positions of the μ-H ligands. In all cases the symmetrical isomer where the μ-H ligand bridges the same Os-Os bond as the sulfur atom (compare Formula II) was formed as the major product, based on ^{1}H NMR, ^{13}C NMR, and IR data [13].

Some compounds listed in Table 6 were prepared by the following methods:

Method I: Compounds of the types $(\mu$-H)Os$_3$(CO)$_9$(μ_3-η^2-EC$_5$H$_4$N) (Formula I; E=O, S, or NR^1) and $(\mu$-H)Os$_3$(CO)$_9$(μ_3-η^2-ECR^2=NR^3) (E=S or NR^1; Formula II) were prepared by thermal decarbonylation of

　　a. $(\mu$-H)Os$_3$(CO)$_{10}$(μ_2-η^2-EC$_5$H$_4$N) and $(\mu$-H)Os$_3$(CO)$_{10}$(μ_2-η^2-NR^{1}CR^{2}NR^3), respectively, (both Section 3.1.1.7.3.2) in refluxing toluene or hydrocarbons between 45 min and 24 h [1, 4, 8, 11], or

　　b. $(\mu$-H)Os$_3$(CO)$_{10}$(μ-SCR^2=NR^3) (see "Organoosmium Compounds" B 3, 1994, pp. 159/61) in refluxing octane or in CH$_2$Cl$_2$ in the presence of 2 equivalents of (CH$_3$)$_3$NO [10, 13, 16]; workup by TLC with CH$_2$Cl$_2$/hexane (1:1) as eluant [16].

　　The reaction conditions are given in Table 6. The products were mostly purified by TLC [4, 8, 10, 11, 13] with CH$_2$Cl$_2$/hexane in various ratios as eluant [10, 11, 13].

Method II: Compounds of the type $(\mu$-H)Os$_3$(CO)$_9$(μ_3-η^2-SCR^2=NR^3) (Formula II) were prepared by photolytic decarbonylation of $(\mu$-H)Os$_3$(CO)$_{10}$(μ-SCR^2=NR^3) (see "Organoosmium Compounds" B 3, 1994, pp. 157/60) upon UV irradiation in hexane for 1 h and workup by TLC with CH$_2$Cl$_2$/hexane (1:9) as eluant [9], or by VIS irradiation in cyclohexane (the conversion is monitored by IR spectroscopy) [16]. Recrystallization from hexane/CH$_2$Cl$_2$ at -20 °C afforded crystalline solids [9, 16]. CO was similarly eliminated on the daylight over a period of days, but not in the dark [2, 9]. $(\mu$-H)Os$_3$(CO)$_9$(μ_3-η^2-SCH=NC$_6$H$_4$F-4) (No. 8) was obtained by such a daylight irradiation at room temperature for 5 d [2].

References on p. 110

(μ-H)Os$_3$(CO)$_9$(μ_3-η^2-NHCH=O) was allegedly formed by thermolysis of (μ-H)Os$_3$(CO)$_{10}$-(μ_2-η^2-NHCH=O) (Section 3.1.1.7.3.2) in refluxing nonane for 40 h [3]; later the product was clearly identified as (μ-H)Os$_3$(CO)$_{10}$(μ_2-η^1-NH$_2$) (Section 3.1.1.7.2.1; 50% yield) by its μ-H resonance in the ^{1}H NMR spectrum [12].

Table 6
Compounds with μ_3-η^2-Bonded N Ligands.
An asterisk preceding the compound number indicates further information at the end of the table, pp. 103/10.
Explanations, abbreviations, and units on p. X.

No. compound	method of preparation (yield) properties and remarks

compound with a μ_3-η^2-bridging ligand bonded by N and O atoms

| 1 (μ-H)Os$_3$(CO)$_9$(μ_3-η^2-OC$_5$H$_4$N)
Formula I | Ia (78% in refluxing toluene for 4 h) [8]; see also [7]
yellow solid [8]
^{1}H NMR (CDCl$_3$): −9.73 (s, OsH); 6.37 (dd, H–3), 6.71 (dt, H–5), 7.51 (dt, H–4), 8.51 (dd, H–6); no values for ^{3}J and ^{4}J given [8]
IR (cyclohexane): 1958 m, 1970 m, 1996 m, 2003 s, 2025 w, 2034 s, 2062 s, 2091 m (all CO) [8]
carbonylation gave (μ-H)Os$_3$(CO)$_{10}$(μ_2-η^2-NC$_5$H$_4$O) (Section 3.1.1.7.3.2); the carbonylation is reversible [7] |

compounds with μ_3-η^2-bridging ligands bonded by N and S atoms

| *2 (μ-H)Os$_3$(CO)$_9$(μ_3-η^2-SC$_5$H$_4$N)
Formula I | Ia (67% in heptane at 160 °C) [4]
from Os$_3$(CO)$_{12}$ and pyridine-2-thiol (ca. 1:3) in toluene at reflux temperature for 11 h; workup by TLC with petroleum ether as eluant (19%), along with 49% of (μ-H)Os$_3$(CO)$_{10}$(μ_2-η^2-NC$_5$H$_4$S) (Section 3.1.1.7.3.2) and 30% of Os(CO)$_2$(SC$_5$H$_4$N)$_2$ (see "Organoosmium Compounds" A 2, 1993, p. 22) [14]
from Os$_3$(CO)$_{10}$(NCCH$_3$)$_2$ (see "Organoosmium Compounds" B 3, 1994, p. 218) and 2,2′-dipyridyl disulfide in CH$_2$Cl$_2$ at room temperature for a few minutes, replacement of the solvent by cyclohexane and heating to reflux temperature for 2 h; workup by extraction of the residue with CH$_2$Cl$_2$ and TLC separation with petroleum ether/CH$_2$Cl$_2$ (85:15) as eluant (5 to 10%), along with No. 3 as the main product and 5 to 10% of Os$_3$(CO)$_{10}$(μ-SC$_5$H$_4$N)$_2$, Os$_2$(CO)$_6$(μ-SC$_5$H$_4$N)$_2$, and |

Table 6 (continued)

No. compound	method of preparation (yield) properties and remarks
	Os(CO)$_2$(SC$_5$H$_4$N)$_2$ (see "Organoosmium Compounds" A 2, 1993, p. 22) [17]

pale-yellow solid [14]; orange crystals containing Os$_3$(CO)$_{10}$(μ-SC$_5$H$_4$N)$_2$ [17]

^{1}H NMR (CDCl$_3$; in the presence of Os$_3$(CO)$_{10}$(μ-SC$_5$H$_4$N)$_2$): −15.17 (OsH), 7.07 (H-5), 7.39 (H-3), 7.54 (H-4), 9.03 (H-6) [17]; (no medium given): −15.10 (s, OsH), 7.04 to 7.65 (m, 3 H of C$_5$H$_4$N), 9.09 (d, 1 H of C$_5$H$_4$N; J=6) [4]

IR (cyclohexane; in the presence of Os$_3$(CO)$_{10}$(μ-SC$_5$H$_4$N)$_2$): 1951 w, 1963 w, 1987 s, 2000 s, 2028 s, 2054 s, 2083 m [17]; (no medium given): 1962 w, 1966 w, 1990 s, 2002 s, 2036 s, 2056 s, 2086 m(d) [4]

a structure with an S-bonded face-capping μ_3-η^1-SC$_5$H$_4$N ligand would be also consistent with the ^{1}H NMR data, but the carbonyl IR patterns make this coordination less plausible [4]

*3 Os$_3$(CO)$_9$(μ_3-η^2-SC$_5$H$_4$N)(μ-SC$_5$H$_4$N)

for formation, see No. 2 [17]

orange crystals [17]

^{1}H NMR (CDCl$_3$): 6.99 (H-5), 7.10 (H-3), 7.15 (H-5′), 7.41 (H-4), 7.53 (H-4′), 7.63 (H-3′), 8.80 (H-6′), 9.02 (H-6) [17]

IR (cyclohexane): 1944 w, 1959 w, 1979 s, 1987 s, 2024 s, 2051 s, 2076 w [17]

*4 Os$_3$(CO)$_9$(μ_3-η^2-N$_2$C$_6$H$_7$S)(μ-SC$_6$H$_7$N$_2$)

by thermolysis of Os$_3$(CO)$_{10}$(μ-SC$_6$H$_7$N$_2$)$_2$ (Formula IV, p. 103) in refluxing n-heptane for 2 h; workup by TLC with CH$_2$Cl$_2$/hexane (2:3) as eluant (42%), along with 9% of No. 5 [18]

from Os$_3$(CO)$_{10}$(μ-SC$_6$H$_7$N$_2$)$_2$ (Formula III, p. 103) and (CH$_3$)$_3$NO (ca. 1:1.5; dropwise addition) in CH$_2$Cl$_2$ at −78 °C, followed by stirring at room temperature for 1 h; workup by TLC with CH$_2$Cl$_2$/hexane (1:4) as eluant (15%), along with 20% of Os$_3$(CO)$_{10}$-(μ-SC$_6$H$_7$N$_2$)$_2$ (Formula IV) and 31% of an uncharacterized orange product [18]

References on p. 110

Table 6 (continued)

No. compound	method of preparation (yield) properties and remarks

*4 (continued)

by carbonylation of No. 5 in CH_2Cl_2 for 3 h
    under 1 atm of CO; workup by TLC with
    CH_2Cl_2/hexane (1:4) as eluant (62%) [18]
pale orange crystals (from n-hexane at
    −20 °C) [18]
^1H NMR (CD_2Cl_2): 2.26, 2.91 (s's, each 1 CH_3),
    2.51 (s, 2 CH_3), 6.86, 6.90 (s's, each 1 CH);
    no further assignment given [18]
IR (CH_2Cl_2): 1936 m, 1954 m, 1981 s, 2023 vs,
    2055 vs, 2078 m [18]
FAB mass spectrum: $[M]^+$ [18]

*5 $Os_3(CO)_8(\mu_3-\eta^2-N_2C_6H_7S)(\mu_2-\eta^2-SC_6H_7N_2)$

by thermolysis of No. 4 in refluxing n-octane
    for 30 min; workup by TLC with CH_2Cl_2/
    hexane (1:1) as eluant (60%); for formation,
    see also No. 4 [18]
deep orange crystals (from CH_2Cl_2/n-hexane
    at −20 °C) [18]
^1H NMR ($CDCl_3$): 2.29, 2.90 (s's, each 1 CH_3),
    2.54 (s, 2 CH_3), 6.89, 7.22 (s's, each 1 CH);
    no further assignments given [18]
IR (n-hexane): 1918 m, 1940 m, 1964 w,
    1980 vs, 1994 m, 2033 vs, 2069 m [18]
FAB mass spectrum: $[M]^+$ [18]

6 $(\mu-H)Os_3(CO)_9(\mu_3-\eta^2-SCH=NCH_3)$
   Formula II

II (63%) [9]
red crystals (from CH_2Cl_2/hexane), m.p. 149.5
    to 150.5 °C [9]
^1H NMR ($CDCl_3$): −14.08 (s, OsH), 3.53 (s,
    CH_3), 10.12 (s, CH) [9]
IR (hexane): 1956 m, 1970 m, 1992 s, 2006 s,
    2032 s, 2058 s, 2089 m (all CO) [9]
thermolysis in refluxing octane for 15 min
    yielded $(\mu-H)Os_3(\mu_2-\eta^2-CH=NCH_3)$-
    $(CO)_9(\mu_3-S)$ (see "Organoosmium Com-
    pounds" B 5, 1994, p. 199; 55%) and
    $(\mu-H)_2Os_6(\mu-CHNCH_3)_2(CO)_{17}(\mu_4-S)(\mu_3-S)$
    ("Organoosmium Compounds" B 9, 1995,
    Section 6.14; 5%) [9]

7 $(\mu-H)Os_3(CO)_9(\mu_3-\eta^2-SCH=NC_6H_5)$
   Formula II

II (61%) [9]
red crystals (from CH_2Cl_2/hexane), m.p. 167 to
    167.5 °C [9]
^1H NMR ($CDCl_3$): −13.69 (s, OsH), 6.90, 7.43
    (m's, 2 H, 3 H of C_6H_5), 10.22 (s, CH) [9]

References on p. 110

Table 6 (continued)

No. compound	method of preparation (yield) properties and remarks
	IR (hexane): 1959 w, 1970 w, 1991 s, 2006 s, 2035 vs, 2056 s, 2089 m (all CO) [9] thermolysis in refluxing octane for 3 h yielded $(\mu\text{-H})Os_3(\mu_2\text{-}\eta^2\text{-CH=NC}_6H_5)(CO)_9(\mu_3\text{-S})$ (see "Organoosmium Compounds" B 5, 1994, p. 199); extended thermolysis gave mainly Os_6 complexes [9] treatment with $P(CH_3)_2C_6H_5$ in CH_2Cl_2/hexane at room temperature for 15 min yielded $(\mu\text{-H})Os_3(CO)_9P(CH_3)_2C_6H_5(\mu_2\text{-}\eta^2\text{-}SCH=NC_6H_5)$ (see "Organoosmium Compounds" B 4b, in preparation) [9]
*8 $(\mu\text{-H})Os_3(CO)_9(\mu_3\text{-}\eta^2\text{-SCH=NC}_6H_4F\text{-}4)$ Formula II	II (69%) [9]; II (45%) [2] red crystals [2]; m.p. 156.5 to 157.5 °C [9] ^{1}H NMR (CDCl$_3$): −13.66 (s, OsH), 6.89, 7.13 (m's, each 2 H of C$_6$H$_4$), 10.24 (s, CH) [2, 9] IR (hexane): 1960 w, 1971 w, 1992 s, 2006 s, 2035 vs, 2058 s, 2089 m (all CO) [2, 9] thermolysis in refluxing octane yielded $(\mu\text{-H})Os_3(\mu_2\text{-}\eta^2\text{-CH=NC}_6H_4F\text{-}4)(CO)_9(\mu_3\text{-S})$ (see "Organoosmium Compounds" B 5, 1994, p. 200) [2, 9]
9 $(\mu\text{-H})Os_3(CO)_9(\mu_3\text{-}\eta^2\text{-SCH=NC}_6H_4CH_3\text{-}4)$ Formula II	II (68%) [9] red crystals (from CH$_2$Cl$_2$/hexane), m.p. 155 to 156 °C [9] ^{1}H NMR (CDCl$_3$): −13.74 (s, OsH), 2.38 (s, CH$_3$), 6.79, 7.20 (d's, each 2 H of C$_6$H$_4$), 10.18 (s, CH) [9] IR (hexane): 1959 w, 1970 w, 1991 s, 2006 s, 2035 vs, 2056 s, 2088 m (all CO) [9] thermolysis in refluxing octane yielded $(\mu\text{-H})Os_3(\mu_2\text{-}\eta^2\text{-CH=NC}_6H_4CH_3\text{-}4)(CO)_9(\mu_3\text{-S})$ (see "Organoosmium Compounds" B 5, 1994, p. 200; 75%) and $(\mu\text{-H})_2Os_6$-$(\mu\text{-CH=NC}_6H_4CH_3\text{-}4)_2(CO)_{17}(\mu_4\text{-S})(\mu_3\text{-S})$ ("Organoosmium Compounds" B 9, 1995, Section 6.14; 5%) [9]
10 $(\mu\text{-H})Os_3(CO)_9\{\mu_3\text{-}\eta^2\text{-SC(NH}_2)=NC_6H_5\}$ Formula II	inseparable mixture of two isomers, based on ^{1}H NMR spectra [16] Ib [16]; II [16] yellow-orange crystals (from CH$_2$Cl$_2$/hexane) [16]

References on p. 110

Table 6 (continued)

No. compound	method of preparation (yield) properties and remarks
10 (continued)	^{1}H NMR (CDCl$_3$): -14.94, -11.83 (both OsH), 5.1 (br, NH$_2$); hydride signals with relative intensities of 0.86:0.14; no resonances for NC$_6$H$_5$ given [16] ^{13}C NMR (CDCl$_3$): 121.9, 127.7, 130.7, 148.8 (C-2,6, C-4, C-3,5 and C-1 of =NC$_6$H$_5$), 183.1 (CNH$_2$), 176.0, 179.5 (each 2 CO at OsS cis to μ-H), 175.7 (2 CO at OsS trans to μ-H), 181.5 (axial CO at OsN), 187.2 (2 equatorial CO at OsN) [16] IR (cyclohexane): 1955 m, 1964 m, 1985 s, 2001 s, 2030 vs, 2053 vs, 2085 m (all CO) [16] mass spectrum: [M]$^+$ [16]
*11 (μ-H)Os$_3$(CO)$_9${μ$_3$-η^2-SC(NHC$_6$H$_5$)=NC$_6$H$_5$} Formula II	inseparable mixture of two isomers, based on ^{1}H NMR spectra [16] Ib [16]; II [16] yellow-orange crystals (from CH$_2$Cl$_2$/hexane) [16] ^{1}H NMR (CDCl$_3$): -14.90, -11.60 (both OsH), 6.32 (NH); hydride signals with relative intensities of 0.86:0.14; no resonances for NC$_6$H$_5$ given [16] ^{13}C NMR (CDCl$_3$, -50 °C): 122.2, 127.9, 131.1, 149.9 (C-2,6, C-4, C-3,5 and C-1 of =NC$_6$H$_5$), 126.3, 128.1, 129.5, 137.4 (C-2,6, C-4, C-3,5 and C-1 of NHC$_6$H$_5$), 183.1 (CNHC$_6$H$_5$), 176.4, 179.5 (each 2 CO at OsS cis to μ-H), 175.5 (2 CO at OsS trans to μ-H), 181.6 (axial CO at OsN), 187.4 (2 equatorial CO at OsN) [16] IR (cyclohexane): 1954 m, 1963 m, 1984 s, 2000 s, 2030 vs, 2052 vs, 2084 m (all CO) [16] mass spectrum: [M]$^+$ [16]
12 (μ-H)Os$_3$(CO)$_9$(μ$_3$-η^2-S$_2$C$_3$H$_4$N) Isomer 1 (E' = S)	inseparable ca. 3:1 mixture of Isomers 1 and 2 differing in the position of the μ-H ligand, based on ^{1}H NMR spectra [13] Ib (69% in refluxing octane for 15 min), along with traces of an isomeric mixture of (μ-H)Os$_3$(μ$_2$-η^2-C$_3$H$_4$NS)(CO)$_9$(μ$_3$-S) (see "Organoosmium Compounds" B 5, 1994, p. 218) [10, 13] orange crystals (from CH$_2$Cl$_2$/hexane) [13] ^{1}H NMR (CDCl$_3$): Isomer 1: -14.41 (OsH; J(Os,H) = 34.1), 3.47 (t, H-4; J(H,H) = 8.3), 3.96 (t, H-5; J(H,H) = 8.3);

References on p. 110

Table 6 (continued)

No. compound	method of preparation (yield) properties and remarks

Isomer 2

Isomer 2: −11.97 (OsH) [13]; see also [10]
^{13}C NMR (CDCl$_3$):
Isomer 1: 37.6 (br, C-4), 70.1 (br, C-5), 198.1 (C-2); 174.9 (CO$_{e,e'}$), 175.5, 178.8 (each 2 CO, CO$_{c,c',d,d'}$), 181.5 (CO$_b$), 187.1 (CO$_{a,a'}$);
Isomer 2: 37.1 (br, C-4), 69.6 (br, C-5), 198.1 (C-2) [13]
IR (cyclohexane): 1954 m, 1960 w, 1968 m, 1991 s, 2004 s, 2009 m, 2031 s, 2034 sh, 2057 s, 2087 m, 2090 w (all CO) [10, 13]
mass spectrum: [M]$^+$ [13]
thermolysis in refluxing octane for 3 h and chromatographic workup gave a mixture of (μ-H)Os$_3$(μ$_2$-η^2-C$_3$H$_4$NS)(CO)$_9$(μ$_3$-S) isomers (66%) which differ in the position of the μ-H ligand spanning the unbridged Os-Os bonds [10, 13]

13 (μ-H)Os$_3$(CO)$_9$(μ$_3$-η^2-SC$_3$H$_5$N$_2$) structure as No. 12, but E′=NH

inseparable ca. 5:4 mixture of Isomers 1 and 2 differing in the position of the μ-H ligand, based on ^{1}H NMR spectra [13]
from Os$_3$(CO)$_{10}$(NCCH$_3$)$_2$ (see "Organoosmium Compounds" B 3, 1994, p. 218) and imidazolidine-2-thione (ca. 1:1) in CH$_2$Cl$_2$ at 40 °C for 5 min, followed by addition of a slight excess of (CH$_3$)$_3$NO and refluxing for 15 min; workup by TLC with CH$_2$Cl$_2$/hexane (3:1) as eluant (26%) [13]
orange-yellow solid (from CH$_2$Cl$_2$/hexane) [13]
^{1}H NMR (CDCl$_3$):
Isomer 1: −14.59 (OsH; J(Os,H)=34.2), 3.2 to 3.6 (m's, H-4,5);
Isomer 2: −12.53 (OsH; J(Os,H)=30.5, 33.5) indicating that the hydride is bridging two nonequivalent Os centers [13]
^{13}C NMR (CDCl$_3$):
Isomers 1 and 2: 48.4 (br, C-4), 60.0 (br, C-5); resonance for C-2 not observed;
Isomer 1: 175.8 (CO$_{e,e'}$), 175.9, 179.3 (each 2 CO, CO$_{c,c',d,d'}$), 182.6 (CO$_b$), 186.3 (CO$_{a,a'}$) [13]
IR (cyclohexane): 1952 m, 1956 w, 1964 m, 1987 s, 2001 s, 2007 m, 2030 s, 2032 sh, 2054 s, 2058 m, 2086 m, 2090 w (all CO), 3461 (νNH) [13]
mass spectrum: [M]$^+$ [13]

References on p. 110

Table 6 (continued)

No. compound	method of preparation (yield) properties and remarks
14 (μ-H)Os$_3$(CO)$_9$(μ$_3$-η^2-SC$_7$H$_4$ON) structure as No. 15, but E′ = O	inseparable ca. 5:3 mixture of Isomers 1 and 2 differing in the position of the μ-H ligand, based on ^{1}H NMR spectra [13] Ib; prepared in refluxing CH$_2$Cl$_2$ for 1.5 h in the presence of 2 equivalents of (CH$_3$)$_3$NO [10, 13] ^{1}H NMR (CDCl$_3$): Isomer 1: −14.30 (OsH; J(Os,H) = 34.4), 7.45 (m, H-4 to H-7); Isomer 2: −12.31 (OsH; J(Os,H) = 30.8, 32.9) indicating that the hydride is bridging two nonequivalent Os centers [13] IR (cyclohexane): 1960 m, 1963 w, 1970 m, 1994 s, 2007 s, 2013 m, 2034 s, 2037 sh, 2059 s, 2062 sh, 2090 w, 2093 w (all CO) [13] mass spectrum: [M]$^+$ [13]
15 (μ-H)Os$_3$(CO)$_9$(μ$_3$-η^2-S$_2$C$_7$H$_4$N) Isomer 1 (E′ = S) Isomer 2	inseparable ca. 7:1 mixture of Isomers 1 and 2 differing in the position of the μ-H ligand, based on ^{1}H NMR spectra [13] Ib (36%); prepared in refluxing CH$_2$Cl$_2$ for 1.5 h in the presence of 2 equivalents of (CH$_3$)$_3$NO; lower yields in refluxing hexane for 24 h [10, 13] orange crystals (from CH$_2$Cl$_2$/hexane) [13] ^{1}H NMR (CDCl$_3$): Isomer 1: −14.59 (OsH; J(Os,H) = 34.2), 7.47 (t, H-5), ca. 7.72 (H-6), ca. 7.73 (H-4), 7.89 (d, H-7; J(H,H) = 8.9); Isomer 2: −12.25 (OsH) [13] ^{13}C NMR (CDCl$_3$): Isomer 1: 122.2 (C-4; J(C,H) = 166.8), 125.5 (C-7; ^{1}J(C,H) = 165.9), 126.6 (C-5; ^{1}J(C,H) = 163.1), 129.0 (C-6; J(C,H) = 163.9), 132.9 (C-9), 151.7 (C-8), 174.2 (CO$_{e,e'}$; ^{2}J(C,μ-H) = 9.2), 174.8, 179.0 (each 2 CO, CO$_{c,c',d,d'}$), 181.3 (CO$_b$), 187.2 (CO$_{a,a'}$), 190.7 (C-2) [13] IR (cyclohexane): 1957 m, 1962 w, 1968 m, 1994 s, 2006 s, 2010 w, 2032 s, 2035 m, 2059 s, 2088 m, 2091 sh (all CO); (hexa-chlorobutadiene and Nujol mulls): μ$_3$-η^2-S$_2$C$_7$H$_4$N bands from 700 to 1600 and intensities given [13] mass spectrum: [M]$^+$ [13]

References on p. 110

Table 6 (continued)

No. compound	method of preparation (yield) properties and remarks

compounds with μ_3-η^2-bridging ligands bonded by two N atoms

*16 $(\mu$-H)Os$_3$(CO)$_9(\mu_3$-η^2-NHC$_5$H$_4$N)
     Formula I

Ia (86% in refluxing cyclohexane for 8 h, 90% in heptane at 68 °C for 5 h) [4, 8]
by thermolysis of $(\mu$-H)Os$_3$(CO)$_9$P(CH$_3)_2$C$_6$H$_5$-$(\mu_2$-η^2-NHC$_5$H$_4$N) (see "Organoosmium Compounds" B 4b, in preparation) in refluxing octane for 2 h under N$_2$, followed by chromatographic workup (ca. 49%), along with equal amounts of $(\mu$-H)Os$_3$-(CO)$_8$P(CH$_3)_2$C$_6$H$_5(\mu_3$-η^2-NHC$_5$H$_4$N) (Section 3.1.1.8.3) [8]
yellow crystals [8]
^{1}H NMR (CDCl$_3$): -12.71 (s, OsH), 4.54 (NH); 6.50 (dd, H–3), 6.66 (dt, H–5), 7.47 (dt, H–4), 8.55 (dd, H–6) [8]
IR (cyclohexane): 1953 m, 1967 m, 1983 m, 1992 w, 2001 s, 2029 s, 2056 s, 2085 m (all CO) [8]
treatment with P(CH$_3)_2$C$_6$H$_5$ in refluxing cyclohexane for 2 min gave $(\mu$-H)Os$_3$(CO)$_9$-P(CH$_3)_2$C$_6$H$_5(\mu_2$-η^2-NHC$_5$H$_4$N) (see "Organoosmium Compounds" B 4b, in preparation) [8]
protonation with CF$_3$CO$_2$H, followed by treatment with KPF$_6$ yielded No. 17; similar protonation with H$_2$SO$_4$ resulted in No. 18, based on ^{1}H NMR spectroscopy [8]
carbonylation gave $(\mu$-H)Os$_3$(CO)$_{10}(\mu_2$-η^2-NC$_5$H$_4$NH) (Section 3.1.1.7.3.2) [7]

17 [$(\mu$-H)$_2$Os$_3$(CO)$_9(\mu_3$-η^2-NHC$_5$H$_4$N)]PF$_6$

by protonation of No. 16 with CF$_3$CO$_2$H and isolation with KPF$_6$ [8]
colorless solid [8]
^{1}H NMR (CDCl$_3$): -14.70, -13.12 (d's, both OsH), 6.94 (dt, H–5), 7.09 (dd, H–3), 7.74 (dt, H–4), 8.36 (dd, H–6) [8]
IR (CH$_2$Cl$_2$): 2013 sh, 2028 m, 2059 sh, 2079 vs, 2105 s, 2138 m (all CO) [8]

18 [$(\mu$-H)$_3$Os$_3$(CO)$_9(\mu_3$-η^2-NHC$_5$H$_4$N)]SO$_4$

by protonation of No. 16 with concentrated H$_2$SO$_4$ (not isolated), based on ^{1}H NMR spectra [8]
pale brown in solution [8]
^{1}H NMR (CDCl$_3$): -16.0 (s, 1 OsH), -14.8 (s, 2 OsH), 7.37 (dd, H–3), 7.58 (dt, H–5), 8.02 (dt, H–4), 8.43 (dd, H–6) [8]

References on p. 110

Table 6 (continued)

No. compound	method of preparation (yield) properties and remarks

19 (μ-H)Os$_3$(CO)$_9$(μ_3-η^2-NHC$_5$H$_3$NCl-6)   by decarbonylation of (μ-H)Os$_3$(CO)$_{10}$(μ_2-η^1-
    Formula I   NHNC$_5$H$_3$Cl-6) (Section 3.1.1.7.2.1) in hexane

                                                 at 130 °C for 14 h under 5 atm of Ar; workup
by TLC with CH$_2$Cl$_2$/hexane (1:1) as eluant
(30%) [5]

^{1}H NMR (acetone-d$_6$): $-$13.00 (s, OsH), 3.31
(br s, NH), 6.82, 7.44 (d's, each 1 H of
C$_5$H$_3$Cl-6; J=8), 7.84 (m, 1 H of C$_5$H$_3$Cl-6)
[5]

IR (hexane): 1958 w, 1968 w, 1983 m, 2002 s,
2031 s, 2057 s, 2086 w (all CO) [5]

mass spectra: [M]$^+$, [M$-$2]$^+$ in a ratio of 1:3
[5]

20 (μ-H)Os$_3$(CO)$_9${μ_3-η^2-N(CH$_2$C$_6$H$_5$)C$_5$H$_4$N}
    Formula I   Ia (36%); obtained in refluxing heptane for
45 min, along with isomeric (μ-H)$_2$Os$_3$-
(μ_2-η^3-C$_6$H$_4$CH$_2$NC$_5$H$_4$N)(CO)$_9$ (see
"Organoosmium Compounds" B 5, 1994,
p. 264) [1, 8]; a reaction time of 100 min or
higher temperatures gave only (μ-H)$_2$Os$_3$-
(μ_2-η^3-C$_6$H$_4$CH$_2$NC$_5$H$_4$N)(CO)$_9$ [8]; see also
[1]

yellow crystals (from hexane) [8]

^{1}H NMR (CDCl$_3$, 29 °C): $-$11.22 (s, OsH), 4.42,
5.40 (d's, each 1 H of CH$_2$; ^{2}J(H,H)=18), 6.23
(dd, H–3), 6.64 (dt, H–4), 7.34 (dt, H–5), 8.66
(dd, H–6), 6.8 to 7.3 (C$_6$H$_5$) [1, 8]; as there is
no symmetry plane passing through the pyridine ring, in contrast to No. 16, the well-
defined AB quartet of the CH$_2$ protons at
$-$36 °C was explained by a μ-H ligand
bridging one of the side Os-Os bonds; with
raising temperature the AB quartet broadens
to coalesce at ca. 25 °C, and gives a sharp
singlet above 40 °C [8]

IR (cyclohexane): 1953 m, 1965 m, 1977 m,
1984 m, 1990 w, 2000 s, 2029 s, 2056 s,
2085 m (all CO) [1, 8]

thermolysis in refluxing heptane resulted in
(μ-H)$_2$Os$_3$(μ_2-η^3-C$_6$H$_4$CH$_2$NC$_5$H$_4$N)(CO)$_9$
("Oganoosmium Compounds" B 5, 1994,
p. 265) [8]

References on p. 110

Table 6 (continued)

No. compound	method of preparation (yield) properties and remarks

21 $(\mu-H)Os_3(CO)_9(\mu_3-\eta^2-NC_4H_3CH=NCH_3)$

from $Os_3(CO)_{10}(NCCH_3)_2$ (see "Organoosmium Compounds" B 3, 1994, p. 218) and $NH(C_4H_3)CH=NCH_3$ (1:1) in refluxing cyclohexane for 20 min; workup by TLC with petroleum ether/ether (1:1) as eluant (50%) [15]
orange-red crystals [15]
^{1}H NMR (CD$_2$Cl$_2$): -14.43 (s, OsH), 3.85 (d, CH$_3$), 6.24 (dd, H-4; J(H-4,5) = 1.9), 6.82 (dd, H-3; J(H-3,4) = 3.9, J(H-3,5) = 1.0), 7.34 (m, CH; J(H,CH$_3$) = 1.0), 7.67 (m, H-5) [15]
IR (cyclohexane): 1951 m, 1964 m, 1982 s, 1995 vs, 2026 s, 2052 vs, 2082 m (all CO) [15]

*22 $(\mu-H)Os_3(CO)_9(\mu_3-\eta^2-NC_4H_3C_4H_6N)$

from $Os_3(CO)_{10}(NCCH_3)_2$ (see "Organoosmium Compounds" B 3, 1994, p. 218) and $HNC_4H_3C_4H_6N$ in refluxing cyclohexane for 1 h; workup as for No. 21 (48%) [15]
orange-red crystals (from hexane) [15]
^{1}H NMR (CD$_2$Cl$_2$): -11.71 (s, OsH), 1.92 (m, H-9), 3.17 (tt, H-10; J(H-9,10) = 8.6, J(H-8,10) = 1.7), 4.36 (tt, H-8; J(H-8,9) = 7.4), 6.19 (dd, H-3; J(H-3,4) = 2.6), 7.08 (dd, H-4: J(H-4,5) = 3.5), 7.33 (dd, H-5; J(H-5,3) = 1.0) [15]
IR (cyclohexane): 1950 m, 1964 m, 1982 m, 1995 vs, 2026 s, 2052 vs, 2082 m (all CO) [15]

23 $(\mu-H)Os_3(CO)_9(\mu_3-\eta^2-C_2H_5N_2)$

inseparable ca. 1:1 mixture of Isomers 1 and 2 [11]
Ia (70% in hexane at 140 °C for 24 h) [11]
^{1}H NMR (CDCl$_3$):
Isomer 1: -14.17 (s, OsH), 1.63 (s, CH$_3$), 4.31, 6.92 (br s's, both NH);
Isomer 2: -11.72 (s, OsH), 1.46 (s, CH$_3$), 3.62 (br s, NH$_2$) [11]
IR (hexane): 1946 m, 1984 sh, 1995 s, 2024 s, 2053 s, 2084 m (all CO) [11]
mass spectrum: [M]$^+$ [11]

Isomer 1

Isomer 2

References on p. 110

Table 6 (continued)

No. compound	method of preparation (yield) properties and remarks

24 $(\mu-H)Os_3(CO)_9\{\mu_3-\eta^2-N(C_3H_7-i)CH=NC_3H_7-i\}$
   Formula II

from $Os_3(CO)_{12}$ and $i-C_3H_7NHCH=NC_3H_7-i$ in refluxing octane for 2.5 h; workup by TLC with ether/pentane (1:1) as eluant (ca. 59%); a reaction time of 9 h gave a yield of only 51%, along with 16% of $Os_2(CO)_6\{\mu-\eta^2-N(C_3H_7-i)CHNC_3H_7-i\}_2$ [1, 7]

by thermolysis of $(\mu-H)Os_3(CO)_9P(CH_3)_2C_6H_5-\{\mu_2-\eta^2-N(C_3H_7-i)CH=NC_3H_7-i\}$ (see "Organoosmium Compounds" B 4b, in preparation) in refluxing octane for 3 h; chromatography as before yielded an inseparable ca. 1:1 mixture with $(\mu-H)Os_3(CO)_8-P(CH_3)_2C_6H_5\{\mu_3-\eta^2-N(C_3H_7-i)CHNC_3H_7-i\}$ (see Section 3.1.1.8.3) [7]

by deprotonation of No. 25 in refluxing CH_2Cl_2 [7

crystals (from hexane) [7]

^1H NMR ($CDCl_3$, 29 °C): -11.46 (s, OsH); 1.03, 1.10 (d's, both CH_3), 2.37, 3.48 (m's, both CH of C_3H_7-i), 9.03 (s, CH=); $^3J(CH_3,CH) = 7$ [1, 7]; variable-temperature ^1H NMR determinations revealed between -60 and $+133$ °C a rigid cluster conformation with two nonequivalent $i-C_3H_7$ groups; at higher temperatures the two initially observed $CH(CH_3)_2$ multiplets were moved together to a single doublet resulting from changes in populations of conformers by rotations about the $N-C_3H_7-i$ bonds [7]

IR (cyclohexane): 1937 vw, 1950 m, 1965 m, 1980 s, 1989 vw, 1999 s, 2027 s, 2054 s, 2084 m (all CO) [1, 7]

reaction with $P(CH_3)_2C_6H_5$ in CH_2Cl_2 at room temperature for 1 h yielded $(\mu-H)Os_3(CO)_9-P(CH_3)_2C_6H_5\{\mu_2-\eta^2-N(C_3H_7-i)CH=NC_3H_7-i\}$ in 87% yield [6, 7]; thermolysis reformed No. 24 by phosphane elimination which is favored over CO elimination [7]; see also Section 3.1.1.8.3 and "Organoosmium Compounds" B 4b, in preparation

protonation with CF_3CO_2H in $CDCl_3$ gave No. 25 [7]

25 $[(\mu-H)_2Os_3(CO)_9\{\mu_3-\eta^2-N(C_3H_7-i)CH=NC_3H_7-i\}][CF_3CO_2]$

by protonation of No. 24 with CF_3CO_2H in $CDCl_3$; oily product (not isolated) [7]

^1H NMR ($CDCl_3$): -13.78, -12.30 (d's, both OsH; J=2), 1.07, 1.15 (d's, each 1 CH_3), 1.36

Table 6 (continued)

No. compound	method of preparation (yield) properties and remarks
	(d, 2 CH$_3$), 2.64, 3.81 (m's, both CH of C$_3$H$_7$-i), 9.30 (s, CH=); ^{3}J(H,H)=7; the nonequivalent i-C$_3$H$_7$ moieties are diastereotopic [7]
	deprotonation in refluxing CH$_2$Cl$_2$ yielded No. 24 [7]
*26　(μ-H)Os$_3$(CO)$_9${μ$_3$-η^2-NHC(C$_6$H$_5$)=NC$_6$H$_5$} 　　　Formula II	Ia (80% in heptane at 97 °C for 6 h under Ar atmosphere) [11]
	from (μ-H)$_2$Os$_3$(CO)$_{10}$ and HN=C(C$_6$H$_5$)NHC$_6$H$_5$ in refluxing toluene at 110 °C for 45 min; workup as before [11]
	^{1}H NMR (CD$_2$Cl$_2$): −11.75 (s, OsH), 4.58 (br s, NH), 6.74 to 7.31 (m, C$_6$H$_5$) [11]
	IR (hexane): 1952 w, 1978 s, 1996 s, 2027 s, 2052 s, 2068 w, 2083 m (all CO) [11]
	mass spectrum: [M]$^+$ [11]

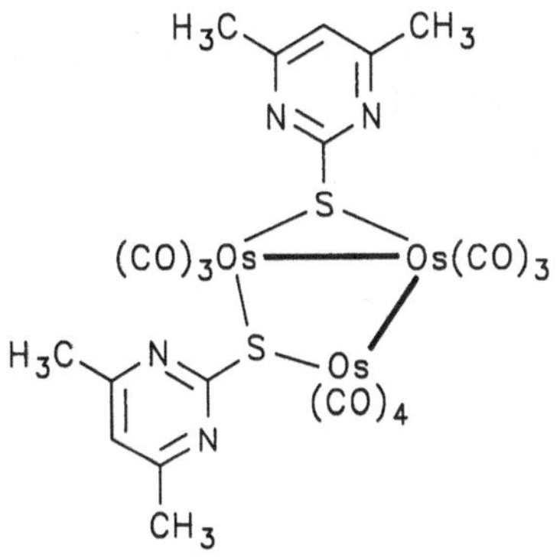

III

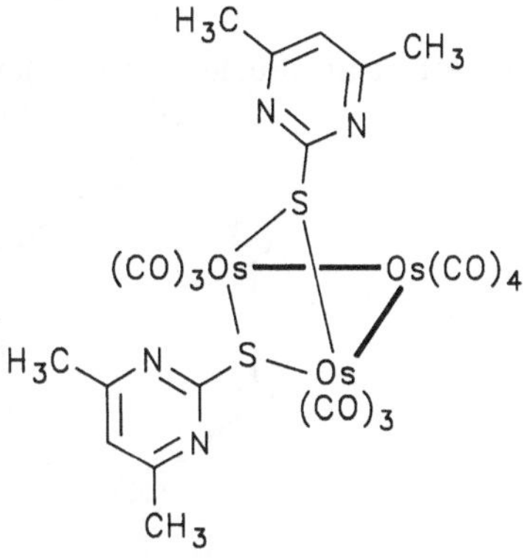

IV

*Further information:

(μ-H)Os$_3$(CO)$_9$(μ$_3$-η^2-SC$_5$H$_4$N) (Table 6, No. 2) crystallizes as a 1:1 mixture with Os$_3$(CO)$_{10}$(μ-SC$_5$H$_4$N)$_2$ in the triclinic space group P$\bar{1}$−C$_i^1$ (No. 2) with a=9.156(3), b=14.970(4), c=17.485(6) Å, α=86.41(2)°, β=75.82(2)°, γ=88.13(2)°; Z=2, D$_c$=2.87 g/cm^3; the structure is shown in **Fig. 35**. The unit cell contains two molecules of the title compound and two molecules of Os$_3$(CO)$_{10}$(μ-SC$_5$H$_4$N)$_2$ [17].

Os$_3$(CO)$_9$(μ$_3$-η^2-SC$_5$H$_4$N)(μ-SC$_5$H$_4$N) (Table 6, No. 3) crystallizes in the triclinic space group P$\bar{1}$−C$_i^1$ (No. 2) with a=8.458(5), b=8.804(2), c=17.181 Å, α=75.95(2)°, β=87.19(4)°, γ=81.81(3)°; Z=2, D$_c$=2.82 g/cm^3; the structure is shown in **Fig. 36**. The cluster consists of an open Os$_3$ framework since the μ$_3$-η^2- and the μ-bonded SC$_5$H$_4$N ligands act as five-electron and three-electron donors, respectively. The nonbonding Os(1)-Os(2) separation amounts to 3.397 Å; the Os(1)-Os(3)-Os(2) angle is 73.4° [17].

References on p. 110

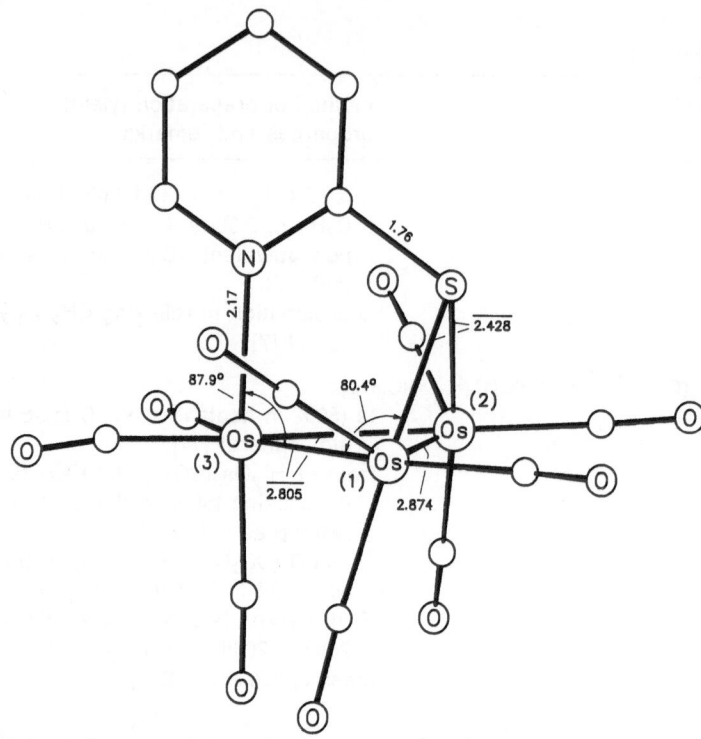

Fig. 35. Molecular structure of $(\mu\text{-H})Os_3(CO)_9(\mu_3\text{-}\eta^2\text{-}SC_5H_4N)$ (No. 2) with selected bond distances (in Å) and bond angles [17].

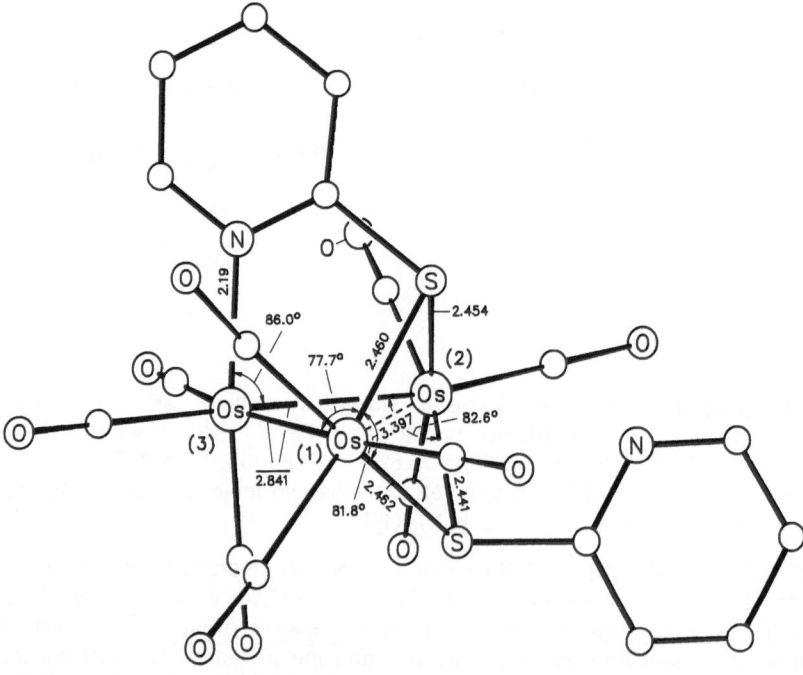

Fig. 36. Molecular structure of $Os_3(CO)_9(\mu_3\text{-}\eta^2\text{-}SC_5H_4N)(\mu\text{-}SC_5H_4N)$ (No. 3) with selected bond distances (in Å) and bond angles [17].

References on p. 110

Os$_3$(CO)$_9$(μ_3-η^2-N$_2$C$_6$H$_7$S)(μ-SC$_6$H$_7$N$_2$) (Table **6**, No. **4**) crystallizes in the orthorhombic space group P2$_1$2$_1$2$_1$ $-$ D$_2^4$ (No. **19**) with a = 10.608(6), b = 20.147(4), c = 13.048(2) Å, β = 90°; Z = 4, D$_c$ = 2.623 g/cm^3. The structure is shown in **Fig. 37**. The molecule consists of an open Os$_3$ framework; the non-bonding Os(1)···Os(3) interactions amount to 3.4281(7) Å. The Os(3)-S(2) distance of 2.463(4) Å is almost 0.03 Å longer than the Os(1)-S(2) bond, indicating the asymmetric bridging nature of the μ_3-η^2-N$_2$C$_6$H$_7$S moiety. The interatomic S···S distance amounts to 3.174(5) Å [18].

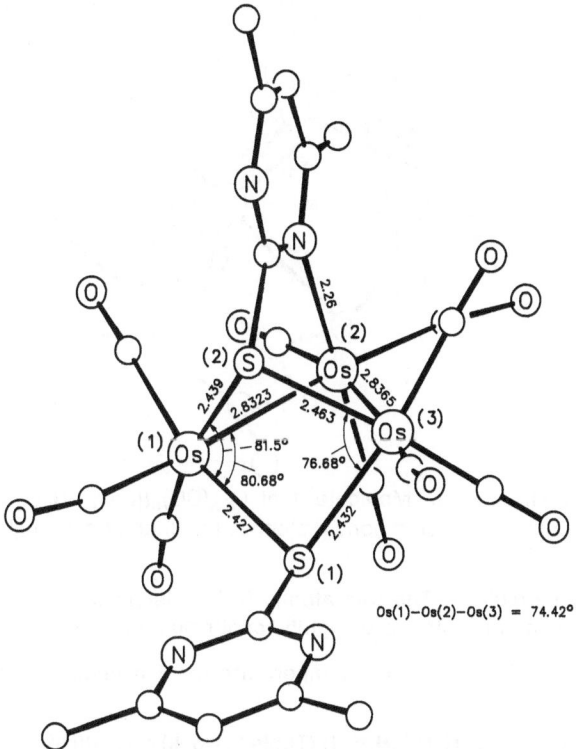

Fig. 37. Molecular structure of Os$_3$(CO)$_9$(μ_3-η^2-N$_2$C$_6$H$_7$S)(μ-SC$_6$H$_7$N$_2$) (No. **4**) with selected bond distances (in Å) and bond angles [18].

Isomerization in CH$_2$Cl$_2$/n-heptane at room temperature for 10 d led to Os$_3$(CO)$_9$(μ_2-η^2-N$_2$C$_6$H$_7$S)(μ-SC$_6$H$_7$N$_2$) (Section 3.1.1.7.3.2) in high yield [18].

Thermolysis in refluxing n-octane for 30 min resulted in No. 5 [18].

Carbonylation in refluxing n-hexane for 3 h under 1 atm of CO yielded Os$_3$(CO)$_{10}$-(μ-SC$_6$H$_7$N$_2$)$_2$ (Formula IV, p. 103) [18].

Os$_3$(CO)$_8$(μ_3-η^2-N$_2$C$_6$H$_7$S)(μ_2-η^2-SC$_6$H$_7$N$_2$) (Table **6**, No. **5**) crystallizes in the monoclinic space group P2$_1$/c $-$ C$_{2h}^5$ (No. **14**) with a = 11.227(4), b = 27.772(3), c = 18.826(3) Å, β = 99.42(1)°; Z = 4, D$_c$ = 2.559 g/cm^3. There are two crystallographically independent molecules in the asymmetric unit. The structure of Molecule 1 is shown in **Fig. 38**. The molecule consists of an open Os$_3$ framework; the non-bonding Os(1)···Os(3) distance amounts to 3.294(1) Å. The Os(3)-S(1) distance of 2.476(4) Å is more than 0.03 Å longer than the Os(1)-S(1) bond, indicating the asymmetric bridging nature of the μ_3-η^2-N$_2$C$_6$H$_7$S moiety. Contrary, the S atom of the μ_2-η^2-N$_2$C$_6$H$_7$S ligand is essentially symmetric coordinated; the two Os-S

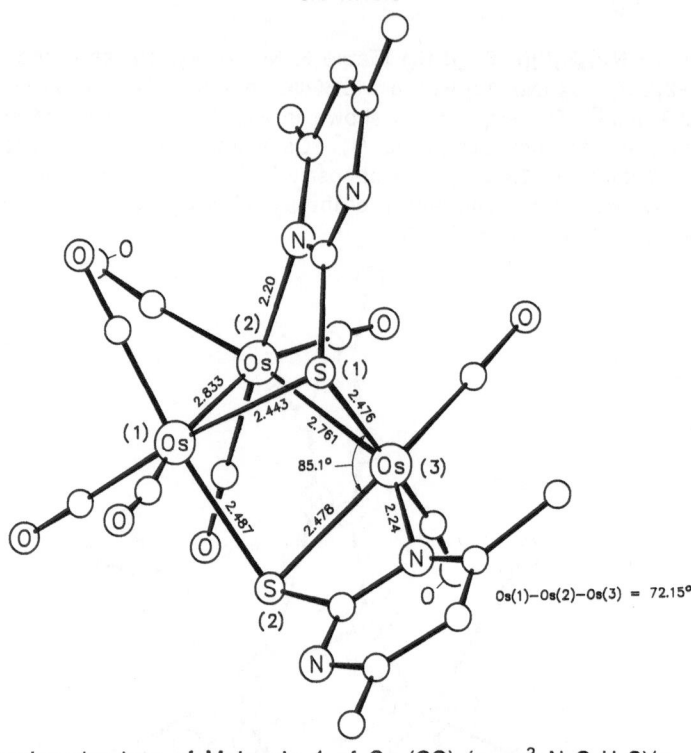

Fig. 38. Molecular structure of Molecule 1 of Os₃(CO)₈(μ₃-η²-N₂C₆H₇S)(μ₂-η²-SC₆H₇N₂)
(No. 5) with selected bond distances (in Å) and bond angles [18].

bonds differ only by 0.009 Å. The interatomic S···S distance amounts to 3.347(7) Å. Most
of the bond lengths in Molecule 2 are slightly longer than in Molecule 1 [18].

Carbonylation in CH_2Cl_2 at room temperature for 3 h under 1 atm of CO gave No. 4
[18].

(μ-H)Os₃(CO)₉(μ₃-η²-SCH=NC₆H₄F-4) (Table 6, No. 8) crystallizes in the monoclinic space
group P2₁/c − C₂h⁵ (No. 14) with a = 12.545(4), b = 10.345(2), c = 17.906(6) Å, β = 110.21(3)°;
Z = 4, D_c = 2.98 g/cm³. The structure is shown in **Fig. 39**. The sulfur atom of the μ₃-η²-
thioformamido ligand bridges the Os(1)–Os(3) bond edge whereas the nitrogen is coordinated to Os(2). The carbon–sulfur bond of 1.774(15) Å is similar to that of a single bond,
while the C–N bond of 1.250(17) Å corresponds to a double bond. The bridging hydride
ligand was not observed crystallographically but it is believed to span the same Os–Os
bond as the sulfur atom but on the opposite side of the Os₃ plane [2, 9].

(μ-H)Os₃(CO)₉{μ₃-η²-SC(NHC₆H₅)=NC₆H₅} (Table 6, No. 11) crystallizes in the monoclinic
space group P2₁/m − C₂h² (No. 11) with a = 9.583(3), b = 14.249(3), c = 9.969(3) Å, β = 103.53(2)°;
Z = 2, D_c = 2.637 g/cm³. The structure is shown in **Fig. 40**. The molecule lies on a crystallo-
graphic mirror plane and hence possesses C_s symmetry. The relatively short Os(1)–Os(2)
bond length of 2.790(1) Å is explained in terms of the geometrical requirements of the
μ₃-η²-SC(NHC₆H₅)=NC₆H₅ ligand as it caps the Os₃ triangle. Contrary, the average Os–S
bonds are elongated by ca. 0.026 Å upon conversion from the μ₂-η²- into the μ₃-η²-bridging
mode. The carbon–sulfur bond of 1.77(1) Å is similar to that of a single bond. The Os–CO
and the C–O bonds range from 1.884(9) to 1.942(8) Å, and from 1.13(11) to 1.17(1) Å, respectively [16].

References on p. 110

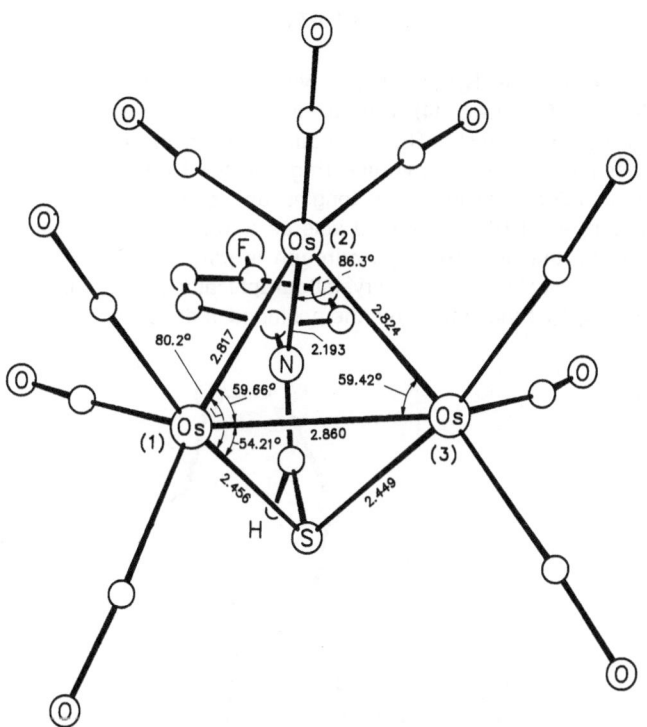

Fig. 39. Molecular structure of $(\mu\text{-H})Os_3(CO)_9(\mu_3\text{-}\eta^2\text{-SCH=NC}_6H_4F\text{-4})$ (No. 8) with selected bond distances (in Å) and bond angles [2, 9].

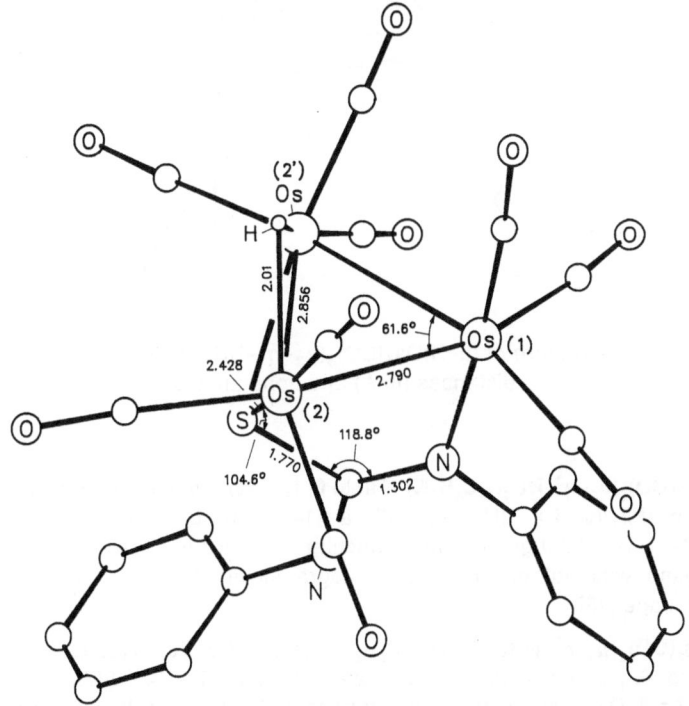

Fig. 40. Molecular structure of $(\mu\text{-H})Os_3(CO)_9\{\mu_3\text{-}\eta^2\text{-SC(NHC}_6H_5)\text{=NC}_6H_5\}$ (No. 11) with selected bond distances (in Å) and bond angles [16].

(μ–H)Os₃(CO)₉(μ₃-η²-NHC₅H₄N) (Table **6**, No. **16**) crystallizes in the monoclinic space group P2₁/n (P2₁/c)−C⁵₂ₕ (No. 14) with a=8.497(2), b=15.019(2), c=15.649(7) Å, β= 103.47(3)°; Z=4, D_c=3.13 g/cm³. The compound, shown in **Fig. 41**, has an approximate mirror plane passing through the pyridine ring and the N(2) and Os(3) atoms, and there is a close correspondence of the bond lengths and angles related by this approximate mirror plane. The μ₃-η²-NHC₅H₄N ligand acts as a five-electron donor; the pyridine ring appears aromatic so that the exocyclic nitrogen is part of a bridging amido group. The metal hydride ligand was not located crystallographically but it is believed to bridge the Os(1)–Os(2) bond edge as indicated from the positions of the carbonyl ligands [8].

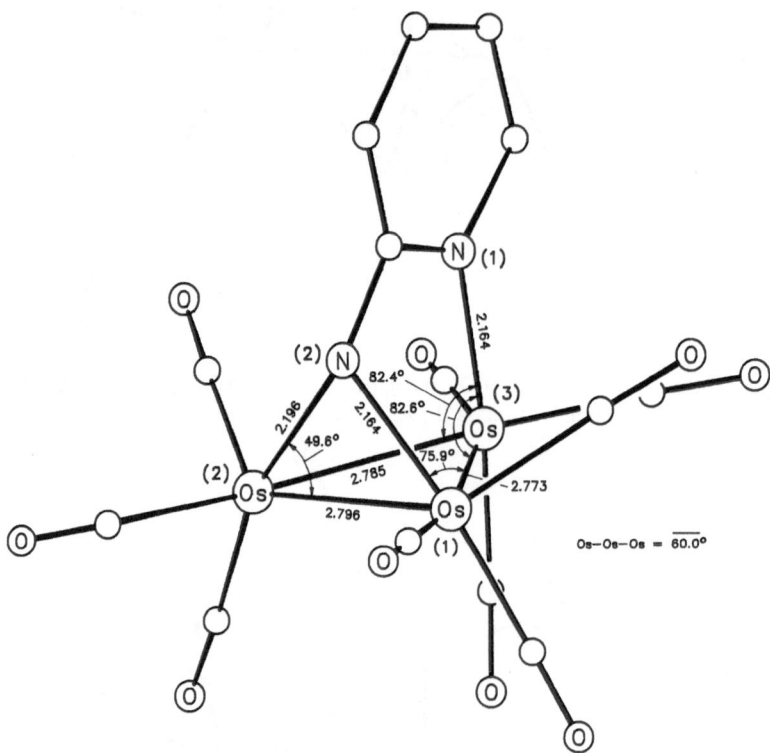

Fig. 41. Molecular structure of (μ–H)Os₃(CO)₉(μ₃-η²-NHC₅H₄N) (No. 16) with selected bond
distances (in Å) and bond angles [8].

(μ–H)Os₃(CO)₉(μ₃-η²-NC₄H₃C₄H₆N) (Table **6**, No. **22**) crystallizes in the monoclinic space group P2₁/n (P2₁/c)−C⁵₂ₕ (No. 14) with a=14.034(3), b=10.661(4), c=14.144(5) Å, β= 91.66(2)°; Z=4, D_c=3.00 g/cm³. The molecular structure is shown in **Fig. 42**. The bridging hydride ligand was not observed crystallographically but it is believed to bridge the Os(1)–Os(2) edge [15].

(μ–H)Os₃(CO)₉{μ₃-η²-NHC(C₆H₅)=NC₆H₅} (Table **6**, No. **26**) crystallizes in the monoclinic space group P2₁/c−C⁵₂ₕ (No. 14) with a=20.002(10), b=14.492(7), c=17.999(6) Å, β= 102.80(3)°; Z=8, D_c=2.66 g/cm³. Two independent but structurally similar molecules, Molecules **A** and **B**, were observed in the asymmetric unit; Molecule **A** is shown in **Fig. 43**.

References on p. 110

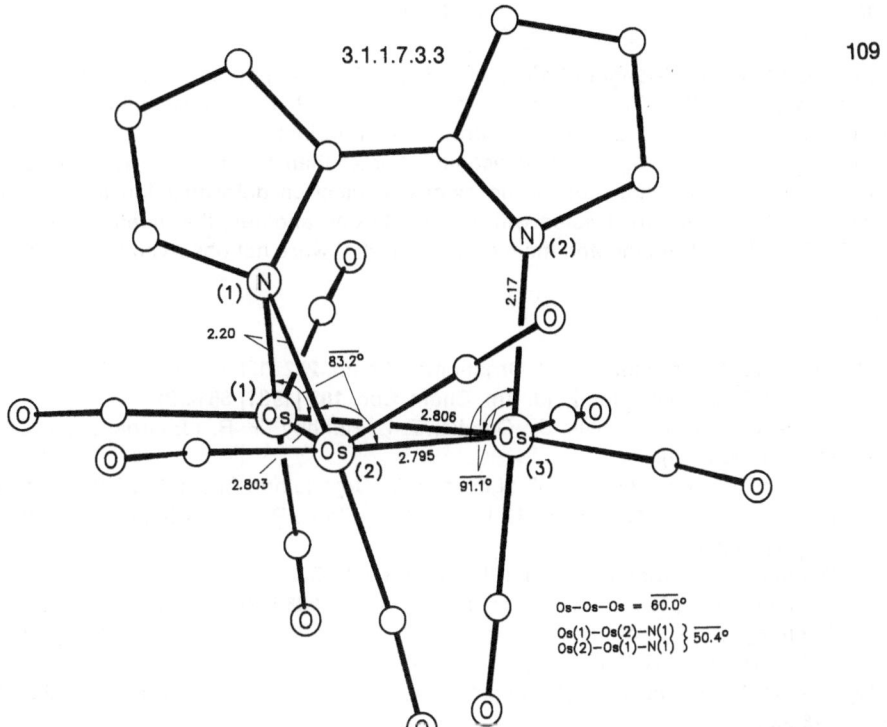

Fig. 42. Molecular structure of (μ-H)Os$_3$(CO)$_9$(μ$_3$-η^2-NC$_4$H$_3$C$_4$H$_6$N) (No. 22) with selected bond distances (in Å) and bond angles [15].

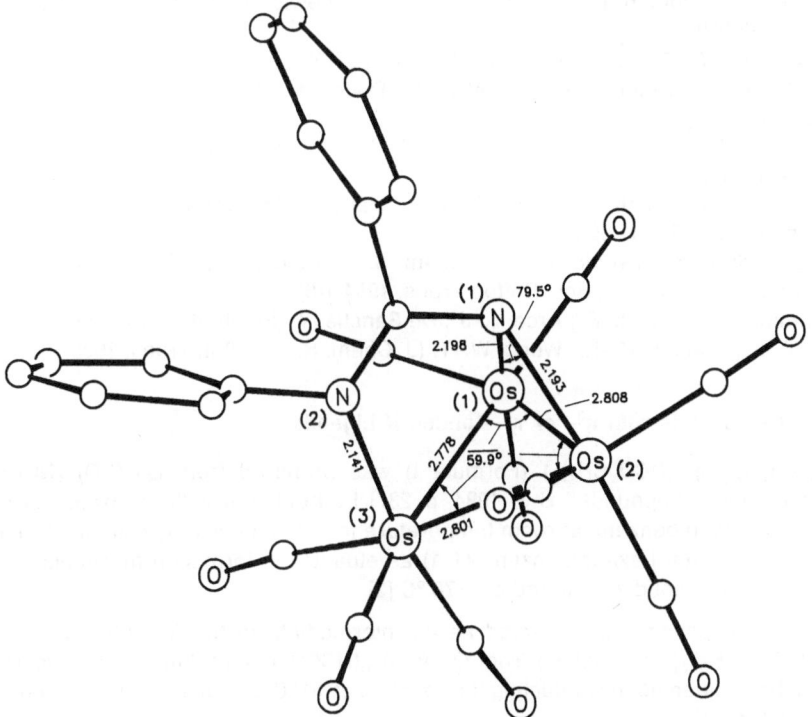

Fig. 43. Molecular structure of Molecule **A** of (μ-H)Os$_3$(CO)$_9${μ$_3$-η^2-NHC(C$_6$H$_5$)=NC$_6$H$_5$} (No. 26) with selected bond distances (in Å) and bond angles [11].

References on p. 110

The equilateral Os_3 triangle of Molecules **A** and **B** is capped by the formally five-electron donating μ_3-η^2-NH=C(C_6H_5)NC_6H_5 unit in the way that N(1) symmetrically spans the Os(1)-Os(2) bond edge and N(2) is terminally coordinated to Os(3) in an axial site. The C-N(2) distance of 1.243 Å is significantly shorter than the average C-N(1) bond length of 1.417 Å indicating that there is no extensive electron delocalization through this bond since the C_6H_5 rings are twisted with respect to one another; the dihedral angle amounts to 56°. Both the μ-H ligand and the hydrogen on N(1) were not observed crystallographically [11].

References:

[1] Deeming, A. J.; Peters, R. (J. Organomet. Chem. **202** [1980] C 39/C 42).

[2] Adams, R. D.; Dawoodi, Z. (J. Am. Chem. Soc. **103** [1981] 6510/2).

[3] Johnson, B. F. G.; Lewis, J.; Odiaka, T. I.; Raithby, P. R. (J. Organomet. Chem. **216** [1981] C 56/C 60).

[4] Burgess, K.; Johnson, B. F. G.; Lewis, J. (J. Organomet. Chem. **233** [1982] C 55/C 58).

[5] Burgess, K.; Johnson, B. F. G.; Lewis, J.; Raithby, P. R. (J. Chem. Soc. Dalton Trans. **1982** 2085/92).

[6] Deeming, A. J.; Manning, P. J. (Philos. Trans. R. Soc. [London] A **308** [1982] 59/66).

[7] Deeming, A. J.; Peters, R. (J. Organomet. Chem. **235** [1982] 221/9).

[8] Deeming, A. J.; Peters, R.; Hursthouse, M. B.; Backer-Dirks, J. D. (J. Chem. Soc. Dalton Trans. **1982** 1205/11).

[9] Adams, R. D.; Dawoodi, Z.; Foust, D. F.; Segmüller, B. E. (Organometallics **2** [1983] 315/23).

[10] Brodie, A. M.; Holden, D. H.; Lewis, J.; Taylor, M. J. (J. Organomet. Chem. **253** [1983] C 1/C 4).

[11] Burgess, K.; Holden, D. H.; Johnson, B. F. G.; Lewis, J. (J. Chem. Soc. Dalton Trans. **1983** 1199/202).

[12] Odiaka, T. I. (J. Organomet. Chem. **284** [1985] 95/9).

[13] Brodie, A. M.; Holden, D. H.; Lewis, J.; Taylor, M. J. (J. Chem. Soc. Dalton Trans. **1986** 633/9).

[14] Deeming, A. J.; Meah, M. N.; Randle, N. P.; Hardcastle, K. I. (J. Chem. Soc. Dalton Trans. **1989** 2211/6).

[15] Arce, A. J.; De Sanctis, Y.; Hernandez, L.; Marquez, M.; Deeming, A. J. (J. Organomet. Chem. **436** [1992] 351/65).

[16] Ainscough, E. W.; Brodie, A. M.; Ingham, S. L.; Kotch, T. G.; Alistair, J. L.; Lewis, J.; Waters, J. M. (J. Chem. Soc. Dalton Trans. **1994** 1/6).

[17] Deeming, A. J.; Vaish, R.; Arce, A. J.; De Sanctis, Y. (Polyhedron **13** [1994] 3285/94).

[18] Au, Y.-K.; Cheung, K.-K.; Wong, W.-T. (J. Chem. Soc. Dalton Trans. **1995** 1047/57).

3.1.1.7.4 Compounds with η^3- or η^4-Bonded N Ligands

$Os_3(CO)_{11}\{\mu_2$-η^3-(NCO$_2$CH$_3$)$_2\}$ (Formula I) was prepared from $Os_3(CO)_{11}$NCCH$_3$ (see "Organoosmium Compounds" B 3, 1994, p. 237) by treatment with an excess of dimethyl azodicarboxylate in benzene at room temperature for 1 h, followed by column chromatography below 15 °C with hexane/benzene (1:1) as eluant; the formation mechanism was discussed. The yellow solid was stored at −78 °C [2].

The title compound was observed as an intermediate in the formation of $Os_3(CO)_{10}$-$\{\mu_2$-η^4-(NCO$_2$CH$_3$)$_2\}$ (see below) from $Os_3(CO)_{10}$(NCCH$_3$)$_2$ and dimethyl azodicarboxylate and could be isolated upon conducting the reaction at 10 °C for 1.5 h, followed by chromatography as before [1].

References on p. 112

^{1}H NMR (CDCl$_3$): 3.64 (s, CH$_3$-1), 3.82 (s, CH$_3$-2) ppm [1, 2].

IR (C$_2$Cl$_4$): 1472 mw, 1530 mw, 1658 mw (CO$_2$CH$_3$-1,2); (C$_6$H$_6$): 2007 s(br), 2020 m, 2040 sh, 2060 s, 2095 s, 2134 mw cm^{-1} (all CO) [1, 2].

Thermolysis in benzene at 70 °C for 10 min led to Os$_3$(CO)$_{10}$\{μ_2-η^4-(NCO$_2$CH$_3$)$_2$\} (see below); a similar conversion was observed in CH$_2$Cl$_2$ at 0 °C within 30 h [2].

Carbonylation in CH$_2$Cl$_2$ at 0 °C for 30 h under 1 atm of CO yielded Os$_3$(CO)$_{12}$\{μ_2-η^2-(NCO$_2$CH$_3$)$_2$\} (Section 3.1.1.7.3.2) [1, 2].

Os$_3$(CO)$_{10}$\{μ_2-η^4-(NCO$_2$CH$_3$)$_2$\} (Formula II) was prepared from Os$_3$(CO)$_{10}$(NCCH$_3$)$_2$ (see "Organoosmium Compounds" B 3, 1994, p. 218) by treatment with an excess of dimethyl azodicarboxylate in benzene at room temperature for 1 h, followed by column chromatography with hexane/benzene (1: 1) as eluant (50 to 87%); yellow, air-stable crystals, m.p. 104 °C (dec., from hexane); the formation of the title compound probably proceeded via Os$_3$(CO)$_{11}$-\{μ_2-η^3-(NCO$_2$CH$_3$)$_2$\} (see above) [1, 2].

The title compound was also obtained by thermolysis of Os$_3$(CO)$_{11}$\{μ_2-η^3-(NCO$_2$CH$_3$)$_2$\} (see above) in benzene at 70 °C for 10 min [1, 2], or from Os$_3$(CO)$_{12}$\{μ_2-η^2-(NCO$_2$CH$_3$)$_2$\} (Section 3.1.1.7.3.2) in benzene at 70 °C for 5 h [2].

Treatment of Os$_3$(CO)$_{12}$\{μ_2-η^2-(NCO$_2$CH$_3$)$_2$\} (Section 3.1.1.7.3.2) with an excess of (CH$_3$)$_3$NO (dissolved in CH$_3$OH) in benzene at room temperature gave Os$_3$(CO)$_{10}$\{μ_2-η^4-(NCO$_2$CH$_3$)$_2$\}, along with Os$_3$(CO)$_{11}$\{μ_2-η^3-(NCO$_2$CH$_3$)$_2$\} as a by-product [2].

^{1}H NMR (CDCl$_3$): 3.86 (s, CH$_3$) ppm [1, 2].

IR (C$_2$Cl$_4$): 1484 w, 1518 w (CO$_2$CH$_3$); (hexane): 1972 w, 1982 w, 1996 w, 2005 vs, 2050 s, 2078 s, 2104 w cm^{-1} (all CO) [1, 2].

FAB mass spectrum (3-mercapto-1,2-propanediol): [M]$^+$ [1, 2].

The title compound crystallizes in the triclinic space group P$\bar{1}$$-C_i^1$ (No. 2) with a= 8.606(1), b=9.412(1), c=14.326(3) Å, α=104.50(1)°, β=97.90(1)°, γ=103.29(1)°; Z=2, D$_c$=3.095 g/cm^3. The structure is shown in **Fig. 44**. The organic ligand spans the non-bonding Os(1)-Os(3) edge forming two five-membered chelate rings with these Os centers to which it is bonded in a tetradentate coordination mode. The N-N distance of 1.439 Å and the N-C distances are close to that of single bonds [1, 2].

Carbonylation in CH$_2$Cl$_2$ for 12 to 24 h under 102 atm of CO gave Os$_3$(CO)$_{12}$\{μ_2-η^2-(NCO$_2$CH$_3$)$_2$\} (Section 3.1.1.7.3.2) in quantitative yield [1, 2].

References on p. 112

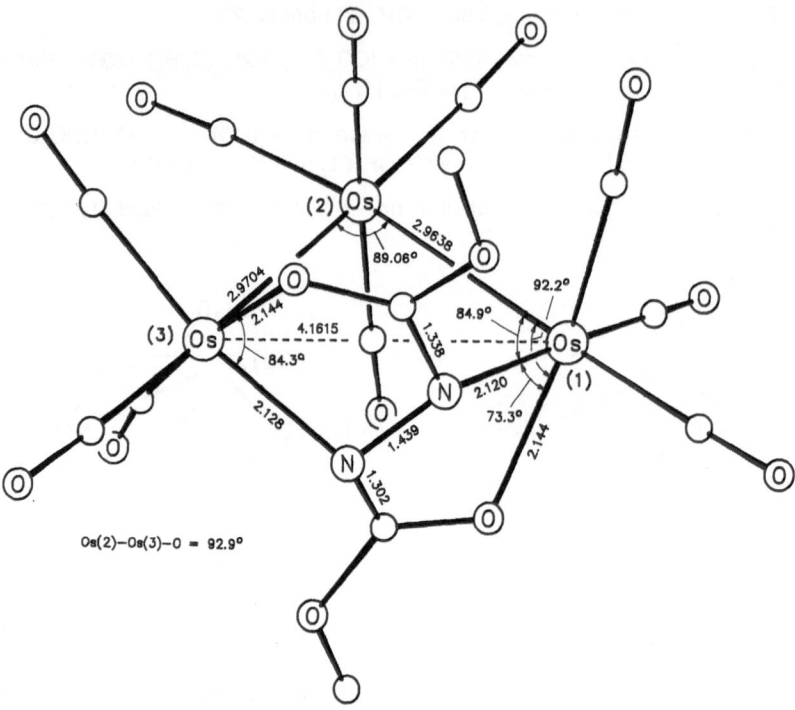

Fig. 44. Molecular structure of $Os_3(CO)_{10}\{\mu_2\text{-}\eta^4\text{-}(NCO_2CH_3)_2\}$ with selected interatomic distances (in Å) and bond angles [1, 2].

References:

[1] Einstein, F. W. B.; Nussbaum, S.; Sutton, D.; Willis, A. C. (Organometallics **2** [1983] 1259/61).
[2] Einstein, F. W. B.; Nussbaum, S.; Sutton, D.; Willis, A. C. (Organometallics **3** [1984] 568/74).

3.1.1.8 Compounds with Additional P-Bonded Ligands

The compounds dealt with in the following Sections **3.1.1.8.1** to **3.1.1.8.3 (present volume)** and **3.1.1.8.4** to **3.1.1.8.6 ("Organoosmium Compounds" B 4b**, in preparation) have an Os_3 framework (mostly triangular), 6 to 11 CO ligands, and at least one additional P-bonded ligand. Other bridging or terminally coordinated halogens and O-, S-, and N-bonded ligands may also be present. Two-electron-donating P ligands (in the following sections designated as ligands of the D type) such as phosphanes and phosphites are generally terminally coordinated, but μ-PR^1R^2 and μ_3-PR ligands were also observed. Bis- and trisphosphanes exhibit mostly a μ-η^2-bridging coordination of the D-D and D-D-D type, but they can also serve as monodentate ligands of the D type. Generally, the P-bonded ligands are coordinated in equatorial positions, essentially lying in the plane of the Os_3 core. The molecular structure of most compounds is based on the triangular $Os_3(CO)_{12}$ framework. Abundantly observed elemental structures are represented by Formulas I to X (hydride ligands omitted). Only a few compounds of the type $Os_3(CO)_{10}D_2Br_2$ (Section 3.1.1.8.5) have a linear Os_3 core (see Formula XI).

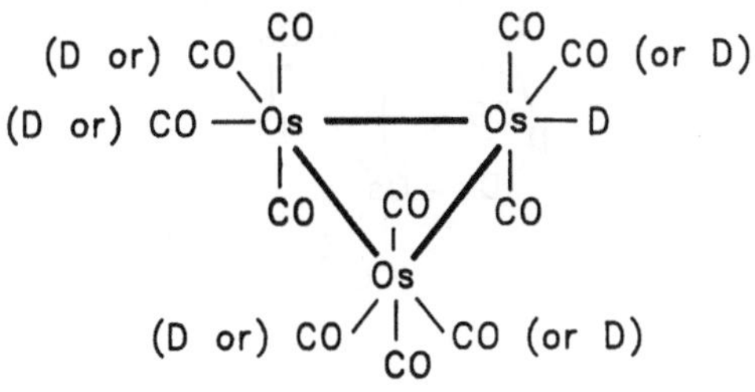

(D or) CO   CO   CO   CO (or D)
(D or) CO—Os ——— Os—D
        CO CO CO
              Os
    (D or) CO   CO (or D)
              CO

I

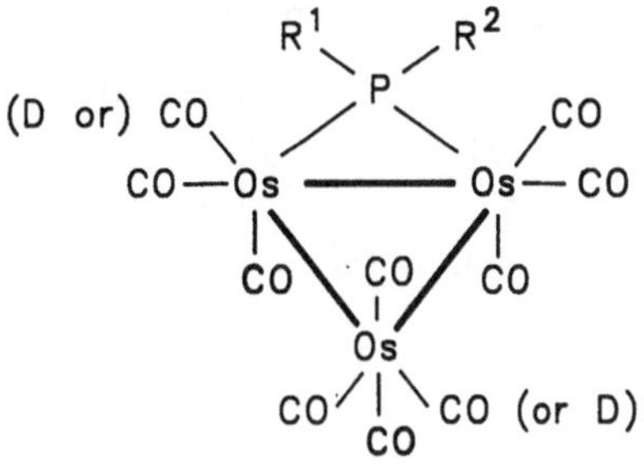

                    R¹   R²
                      P
(D or) CO   CO
    CO—Os ——— Os—CO
        CO CO CO
              Os
        CO   CO (or D)
              CO

II

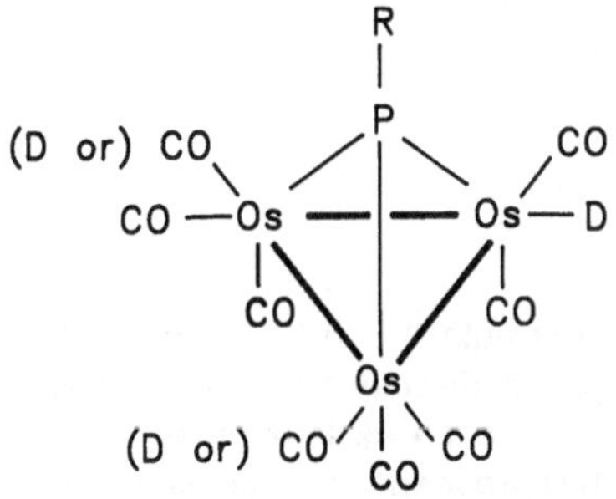

              R
              P
(D or) CO   CO
    CO—Os ——— Os—D
        CO   CO
              Os
    (D or) CO   CO
              CO

III

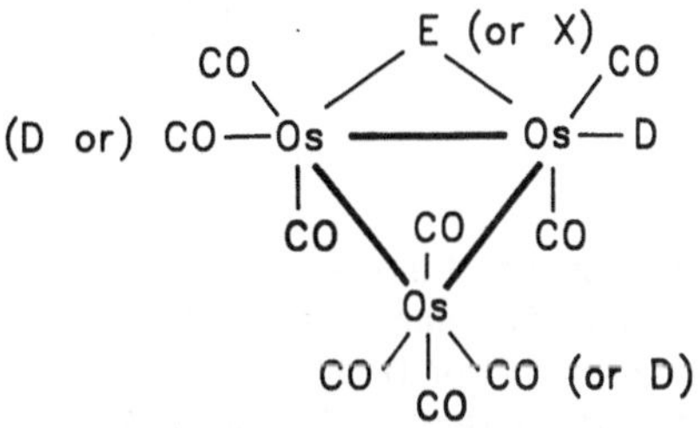

                E (or X)
        CO   CO
(D or) CO—Os ——— Os—D
        CO CO CO
              Os
        CO   CO (or D)
              CO

IV

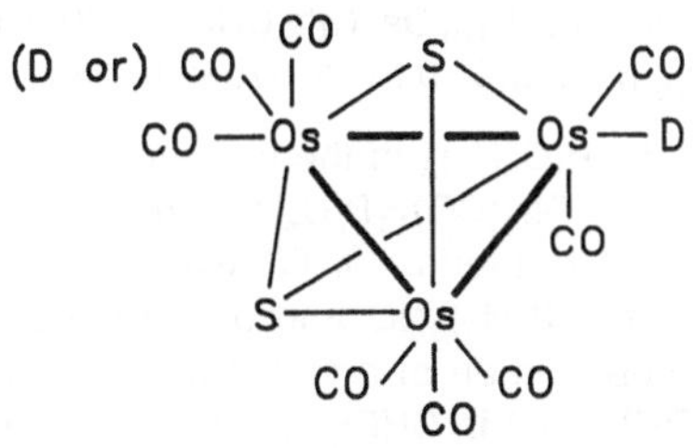

(D or) CO   CO   S   CO
    CO—Os ——— Os—D
        CO   CO
        S ——— Os
        CO   CO
              CO

V

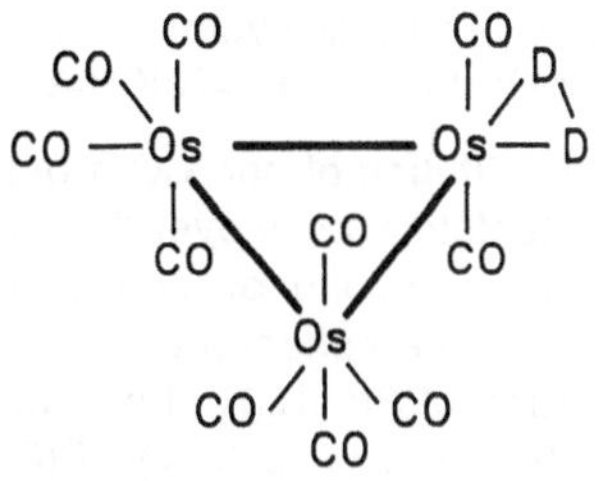

        CO   CO   CO   CO
    CO—Os ——— Os—D
        CO CO CO
              Os
        CO   CO
              CO

VI

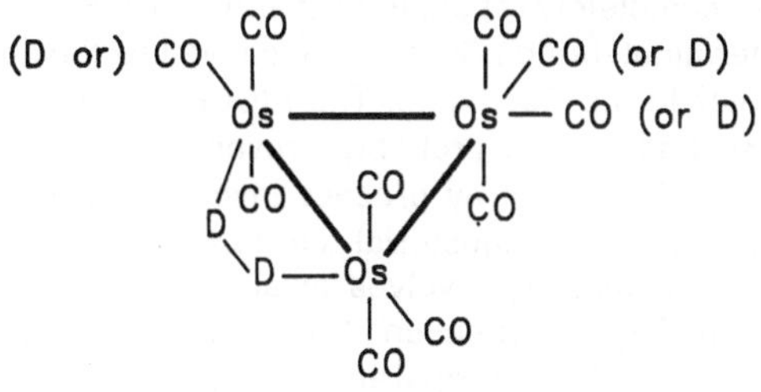

(D or) CO   CO   CO   CO (or D)
    Os ——— Os—CO (or D)
        CO CO CO
        D—Os
        CO   CO
              CO

VII

        CO   CO   CO   CO
    CO—Os ——— Os—CO
        CO CO CO
    D   D—Os—D   D
              CO

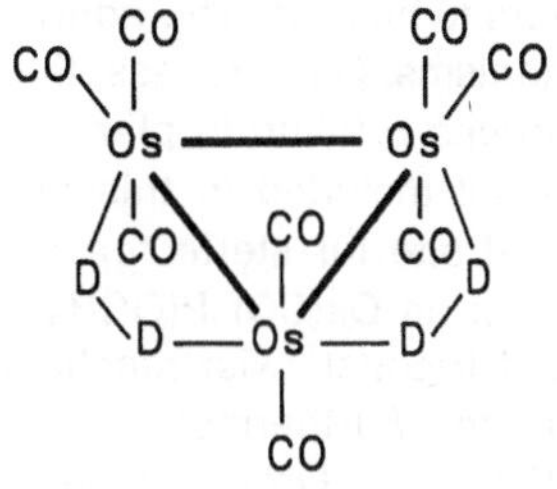

VIII

References on pp. 115/6

CO—Os———Os—CO with E (or X) bridge, D—Os, etc.

IX

CO—Os———Os—CO with S bridge, D—Os, etc.

X

$$D-Os-----Os-----Os-D$$
(CO, CO, Br; CO, CO; Br, CO, CO substituents)

XI

Compounds of the types $Os_3(CO)_{12-n}D_n$ ($n=1$ to 4; Formula I) and $Os_3(CO)_{12-n}(D-D)_n$ ($n=1$ or 2; Formulas VII and VIII) were typically prepared directly from $Os_3(CO)_{12}$ and an excess of D or D-D, respectively, in a suitable solvent such as toluene or xylene at elevated temperatures [1, 13 to 16, 18]. The reactions with D proceed via successive substitution of CO by D to give mixtures of $Os_3(CO)_{11}D$, $Os_3(CO)_{10}D_2$, $Os_3(CO)_9D_3$, and $Os_3(CO)_8D_4$ [15, 16, 18]. $Os_3(CO)_7D_5$ and $Os_3(CO)_6D_6$ were better obtained from $Os_3(CO)_8D_4$ by UV irradiation in the presence of D [8, 16]. Similar preparations can also be performed with $Os_3(CO)_{12}$ under mild conditions in the presence of $(CH_3)_3NO$ as a CO labilization agent [6, 19, 20] or by displacement of the labile groups in $Os_3(CO)_{12-n}(NCCH_3)_n$ ($n=1$ or 2), $Os_3(CO)_{11}(\eta^1-OC_4H_8-cyclo)$, $Os_3(\eta^2-1,5-C_8H_{14}-cyclo)(CO)_{10}$, $Os_3(\eta^4-CH_2=CHCH=CH_2-cis)$-$(CO)_{10}$, or $Os_3(\mu-\eta^4-CH_2=CHCH=CH_2-trans)(CO)_{10}$ [2, 3, 5, 7, 9 to 12, 17, 22 to 24].

The substitution of one CO group of $Os_3(CO)_{12}$ by $P(C_6H_5)_3$ in the presence of $(CH_3)_3NO$ (both in a tenfold excess over the metal cluster) to give $Os_3(CO)_{11}P(C_6H_5)_3$ in $CHCl_3/C_2H_5OH$ (2:1) obeys a second-order rate law. The rates of formation of $Os_3(CO)_{11}P(C_6H_5)_3$ are first-order in the concentrations of metal cluster and $(CH_3)_3NO$ and of zero-order in the concentration of $P(C_6H_5)_3$ [19]. Similarly, further substitution of a CO group of $Os_3(CO)_{11}D$ by D ($D=P(OCH_3)_3$, $P(OC_2H_5)_3$, $P(C_4H_9-n)_3$, and $P(C_6H_5)_3$) in $CHCl_3$ leading to $Os_3(CO)_{10}D_2$ obeys the same rate law as before under pseudo-first-order conditions [20].

Compounds of the type $Os_3(CO)_{12-n}\{P(OCH_3)_3\}_n$ ($n=5$ to 1; Sections 3.1.1.8.2 to 3.1.1.8.6) are stereochemically nonrigid in solution, while $Os_3(CO)_6\{P(OCH_3)_3\}_6$ (Section 3.1.1.8.1) appears to be rigid. The nonrigidity can be completely rationalized in terms of two exchange mechanisms. One process, the pairwise terminal-bridge CO exchange involves the six axial CO groups rotating in planes perpendicular to the Os_3 plane. The $P(OCH_3)_3$ groups seem to be not included in this process; these ligands are probably prevented from entering an axial site for steric reasons. A second, higher-energy process involves the $P(OCH_3)_3$ ligand of an $Os(CO)_3P(OCH_3)_3$ unit moving from one equatorial site to the other via a restricted trigonal-twist mechanism; this simultaneously involves axial-equatorial carbonyl exchange. Furthermore, for $Os_3(CO)_8\{P(OCH_3)_3\}_4$ (Section 3.1.1.8.3) and $Os_3(CO)_{10}$-$\{P(OCH_3)_3\}_2$ ("Organoosmium Compounds" B 4b, in preparation) rapid, nondissociative equilibria of isomers have been observed [8, 15].

Treatment of $Os_3(CO)_{12}$ with primary and secondary phosphanes led to μ-PR^1R^2-bridged hydrido clusters (Formula II) in toluene at ca. 100 °C, while under severe conditions (refluxing nonane) μ_3-PR species (Formula III) were formed [4, 6, 21]. These reactions proceeded via phosphane adducts (Formula I) which rearrange by hydride migration and further loss of CO. Similarly, reactions of $Os_3(CO)_{12}$ with PH_2R in the presence of $(CH_3)_3NO$ at room temperature give $Os_3(CO)_{11}PH_2R$ and $(\mu$-$H)Os_3(CO)_{10}(\mu$-PHR) but at higher temperatures $(\mu$-$H)_2Os_3(CO)_9(\mu_3$-PR); $R = C_6H_5$, $C_6H_4CH_3$-4, cyclo-C_6H_{11} [4, 6].

If not stated otherwise, the ^{31}P NMR spectra are referenced to 85% H_3PO_4, and the δ values having a positive sign for downfield shifts. In many cases, the spectra are referenced versus internal $P(OCH_3)_3$ (ca. +140 ppm versus 85% H_3PO_4), generally leading to negative δ values.

References:

[1] Crow, J. P.; Cullen, W. R. (Inorg. Chem. **10** [1971] 1529/31).

[2] Tachikawa, M.; Richter, S. I.; Shapley, J. R. (J. Organomet. Chem. **128** [1977] C 9/C 14).

[3] Johnson, B. F. G.; Lewis, J.; Pippard, D. A. (J. Organomet. Chem. **145** [1978] C 4/C 6).

[4] Iwasaki, F.; Mays, M. J.; Raithby, P. R.; Taylor, P. L.; Wheatley, P. J. (J. Organomet. Chem. **213** [1981] 185/206).

[5] Johnson, B. F. G.; Lewis, J.; Pippard, D. A. (J. Chem. Soc. Dalton Trans. **1981** 407/12).

[6] Natarajan, K.; Zsolnai, L.; Huttner, G. (J. Organomet. Chem. **220** [1981] 365/81).

[7] Deeming, A. J.; Donovan-Mtunzi, S.; Kabir, S. E. (J. Organomet. Chem. **276** [1984] C 65/C 68).

[8] Alex, R. F.; Pomeroy, R. K. (J. Organomet. Chem. **284** [1985] 379/84).

[9] Deeming, A. J.; Donovan-Mtunzi, S.; Kabir, S. E.; Manning, P. J. (J. Chem. Soc. Dalton Trans. **1985** 1037/41).

[10] Patel, V. D.; Cherkas, A. A.; Nucciarone, D.; Taylor, N. J.; Carty, A. J. (Organometallics **4** [1985] 1792/800).

[11] Shojaie, A.; Atwood, J. D. (Organometallics **4** [1985] 187/90).

[12] Bruce, M. I.; Horn, E.; Bin Shawkataly, O.; Snow, M. R.; Tiekink, E. R. T.; Williams, M. L. (J. Organomet. Chem. **316** [1986] 187/211).

[13] Cartwright, S.; Clucas, J. A.; Dawson, R. H.; Foster, D. F.; Harding, M. M.; Smith, A. K. (J. Organomet. Chem. **302** [1986] 403/12).

[14] Clucas, J. A.; Dawson, R. H.; Dolby, P. A.; Harding, M. M.; Pearson, K.; Smith, A. K. (J. Organomet. Chem. **311** [1986] 153/62).

[15] Alex, R. F.; Pomeroy, R. K. (Organometallics **6** [1987] 2437/46).

[16] Alex, R. F.; Einstein, F. W. B.; Jones, R. H.; Pomeroy, R. K. (Inorg. Chem. **26** [1987] 3175/8).

[17] Deeming, A. J.; Donovan-Mtunzi, S.; Kabir, S. E. (J. Organomet. Chem. **333** [1987] 253/62).

[18] Bruce, M. I.; Liddell, M. J.; Hughes, C. A.; Patrick, J. M.; Skelton, B. W.; White, A. H. (J. Organomet. Chem. **347** [1988] 181/205).

[19] Shen, J. K.; Shi, Y. L.; Gao, Y. C.; Shi, Q. Z.; Basolo, F. (J. Am. Chem. Soc. **110** [1988] 2414/8).

[20] Shen, J. K.; Gao, Y. C.; Shi, Q. Z.; Basolo, F. (Inorg. Chem. **27** [1988] 4236/9).

[21] Arif, A. M.; Bright, T. A.; Heaton, D. E.; Jones, R. A.; Nunn, C. M. (Polyhedron **9** [1990] 1573/87).

[22] Ang, H. G.; Cai, Y. M.; Kwik, W. L.; Morrison, E. C.; Tocher, D. A. (J. Organomet. Chem. **403** [1991] 383/5).

[23] Cullen, W. R.; Rettig, S. J.; Zheng, T. C. (Can. J. Chem. **70** [1992] 2215/23).

[24] Ang, H. G.; Koh, C. H.; Koh, L. L.; Kwik, W. L.; Leong, W. K.; Leong, W. Y. (J. Chem. Soc. Dalton Trans. **1993** 847/55).

3.1.1.8.1 Compounds with Six Carbonyl Groups

The compounds in Table 7 are of the type $Os_3(CO)_6D_5D'$ and $(\mu\text{-H})_2Os_3(CO)_6(\mu\text{-PR}_2)_2D$; D,D' is phosphite or phosphane. The structure of the latter compound is figured in Table 7.

Table 7
Compounds with Six Carbonyl Groups.
An asterisk preceding the compound number indicates further information at the end of the table, pp. 117/9.
Explanations, abbreviations, and units on p. X.

No. compound	method of preparation (yield) properties and remarks
*1 $Os_3(CO)_6\{P(OCH_3)_3\}_6$	by UV irradiation (Hanovia 200-W lamp) of $Os_3(CO)_8\{P(OCH_3)_3\}_4$ (Section 3.1.1.8.3; in situ prepared) in heptane for 48 h in the presence of an excess of $P(OCH_3)_3$; workup by washing of the precipitate with hexane and hexane/toluene (31%) [3]; similar UV irradiation but only for 8 h yielded only minor amounts of No. 1 along with $Os_3(CO)_7\{P(OCH_3)_3\}_5$ (Section 3.1.1.8.2) as the main product [1]
	red–orange prisms (from toluene or CH_2Cl_2/hexane); m.p. 166.5 to 168 °C (dec.) [3]
	^1H NMR (CD_2Cl_2 or $CDCl_3$): 3.60 (br s, CH_3); (CD_2Cl_2, -50 and -54 °C): 3.53 (d, CH_3; $J(P,H)=11.2$); (toluene-d_8, $+70$ °C): 3.61 (d, CH_3; $J(P,H)=10.9$) [3]
	^{13}C $\{^1H\}$ NMR (CH_2Cl_2/CD_2Cl_2, -57 °C): 203 (t, CO; $J(P,C)=12.0$) [3]
	^{31}P $\{^1H\}$ NMR (CD_2Cl_2 or $CDCl_3$): -42.6 (br s, peak width half-heights of 17.5 Hz at ca. 0.7×10^{-3} M and 28 Hz at ca. 7×10^{-3} M); (CD_2Cl_2, -50 °C): -41.4 (s); (toluene/toluene-d_8, -89 °C): -39.8 (s); (toluene-d_8, $+70$ °C): -43.04 (s) [3]
	UV (CH_2Cl_2): 335, 412, 475 [3]
	IR (CH_2Cl_2): 2013, 1938, 1881 (all CO) [1, 3]
	mass spectrum: $[M]^+$, $[M-P(OCH_3)_3-2\,CH_3]^+$ [3]
	treatment with $P(OC_2H_5)_3$ in CH_2Cl_2/CD_2Cl_2 at room temperature or in toluene-d_8 at 70 °C for 1 h did not lead to an incorporation of $P(OC_2H_5)_3$, based on ^1H and ^{31}P NMR spectra [3]

Table 7 (continued)

No. compound	method of preparation (yield) properties and remarks

2   $Os_3(CO)_6\{P(OCH_3)_3\}_5P(OC_2H_5)_3$ | by UV irradiation of $Os_3(CO)_7\{P(OCH_3)_3\}_5$ (Section 3.1.1.8.2) in hexane in the presence of an excess of $P(OC_2H_5)_3$ (low yields) [2]

*3 $(\mu\text{-H})_2Os_3(CO)_6\{\mu\text{-P}(C_4H_9\text{-t})_2\}_2PH(C_4H_9\text{-t})_2$

from $Os_3(CO)_{12}$ and $PH(C_4H_9\text{-t})_2$ (1:3) in refluxing toluene for 24 h; workup by column chromatography on alumina with hexane as eluant (15.4%), along with 50% of $(\mu\text{-H})_2Os_3(CO)_8\{\mu\text{-P}(C_4H_9\text{-t})_2\}_2$ (Section 3.1.1.8.3) [4]

from $(\mu\text{-H})_2Os_3(CO)_8\{\mu\text{-P}(C_4H_9\text{-t})_2\}_2$ and an excess of $PH(C_4H_9\text{-t})_2$ in refluxing THF for 24 h [4]

red, air-stable crystals (from hexane by slow cooling to $-10\ ^\circ$C); m.p. 178 to 182 $^\circ$C [4]

^{1}H NMR (C_6D_6): 5.65 (dd, PH; ^{1}J(P,H)=339.0, ^{3}J(P,H)=9.61), 1.61 (d, CH_3 of $PH(C_4H_9\text{-t})_2$; ^{2}J(P,H)=14.3), 1.36, 1.09 (d's, CH_3 of $\mu\text{-P}(C_4H_9\text{-t})_2$; ^{2}J(P,H)=14.3, 14.6), -9.76, -11.72 (m's, OsH; ^{2}J(P,H)=10.5, 10.3) [4]

^{31}P $\{^1$H$\}$ NMR (C_6D_6): 183.07 (d, μ-P; ^{2}J(P,P)=88.5), 146.67 (dd, μ-P; ^{2}J(P,P)=92.7), 29.10 (br t, PH; ^{2}J(P,P)=15.5) [4]

^{31}P NMR (C_6D_6): 29.10 (dt, PH; ^{1}J(P,H)=333.8) [4]

resonances and coupling constants for the ^{31}P $\{^1$H$\}$ and ^{31}P NMR spectra given in the experimental and the text part of [4] differ slightly

IR (Nujol mull): 2347, 2010, 1971, 1938, 1915, 1391, 1310, 1261, 1169, 1096, 1023, 936, 889, 862, 843, 806, 598, 573, 549 (no assignments given) [4]

*Further information:

$Os_3(CO)_6\{P(OCH_3)_3\}_6$ (Table 7, No. 1) crystallizes in the rhombohedral space group $R\bar{3}-C_{3i}^2$ (No. 148) with a=12.898(9), c=22.705(15) Å; Z=9, D_c=2.259 g/cm^3 at $-60\ ^\circ$C. The structure is shown in **Fig. 45**. The phosphite ligands occupy the six less sterically crowded equatorial sites of the Os_3 triangle. The $Os(CO)_2\{P(OCH_3)_3\}_2$ groups are twisted to one another, so that the molecule has D_3 symmetry [3].

$(\mu\text{-H})_2Os_3(CO)_6\{\mu\text{-P}(C_4H_9\text{-t})_2\}_2PH(C_4H_9\text{-t})_2$ (Table 7, No. 3) crystallizes in the monoclinic space group $P2_1/c-C_{2h}^5$ (No. 14) with a=20.689(7), b=11.507(4), c=17.102(6) Å, $\alpha=90^\circ$, $\beta=$

References on p. 119

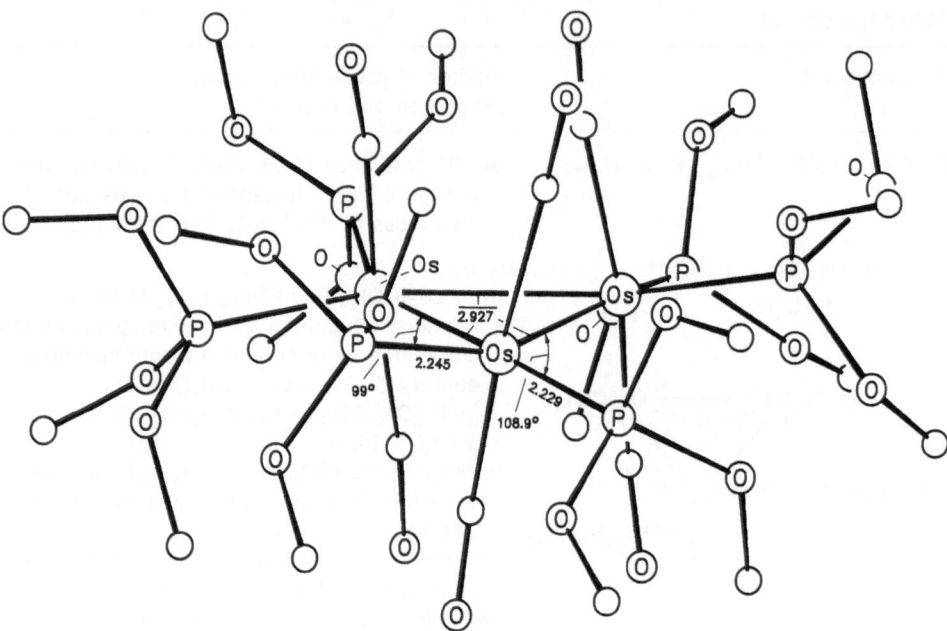

Fig. 45. Molecular structure of $Os_3(CO)_6\{P(OCH_3)_3\}_6$ (No. 1) with selected bond distances (in Å) and bond angles [3].

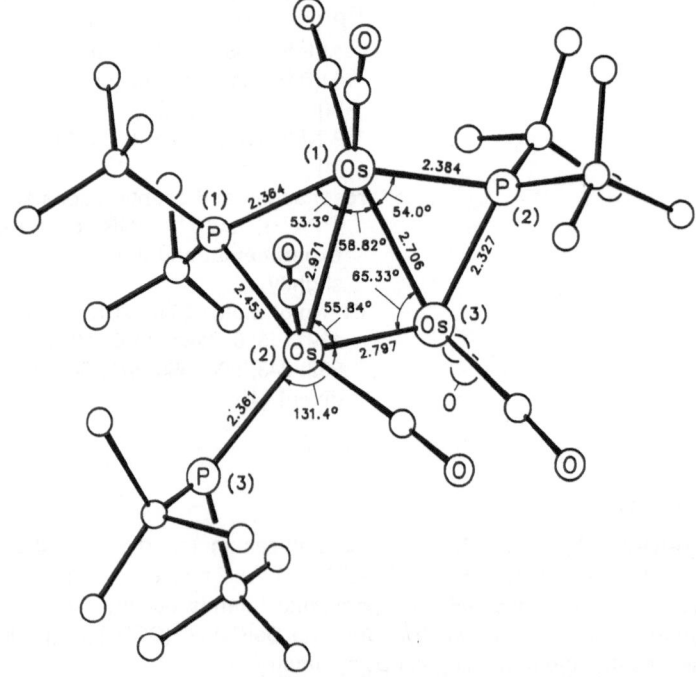

Fig. 46. Molecular structure of $(\mu\text{-H})_2Os_3(CO)_6\{\mu\text{-P}(C_4H_9\text{-t})_2\}_2PH(C_4H_9\text{-t})_2$ (No. 3) with selected bond distances (in Å) and bond angles [4].

$109.61(3)°$, $\gamma = 90°$; $Z = 4$, $D_c = 2.039$ g/cm^3. The structure of the 46-electron molecule is shown in **Fig. 46**. The coordination geometries of Os(1) and Os(2) are roughly octahedral upon ignoring the metal–metal interactions, but the coordination geometry of Os(3) is trigonal-bipyramidal. The hydride ligands were not located crystallographically, but they are believed to bridge the Os(1)–Os(3) and Os(2)–Os(3) bonds. Although even the spectroscopic data do not permit an unequivocal assignment of their positions they are likely bonded to the coordinatively and (formally) electronically unsaturated Os(3) center [4].

Thermolysis in n-butyl ether at 140 °C for 24 h yielded $(\mu\text{-H})_2Os_3(CO)_7(\mu_3\text{-PC}_4H_9\text{-t})\text{-}$ $\{PH(C_4H_9\text{-t})_2\}_2$ (Section 3.1.1.8.2) [4].

Carbonylation in hexane under 1 atm of CO gave $(\mu\text{-H})_2Os_3(CO)_8\{\mu\text{-P}(C_4H_9\text{-t})_2\}_2$ (Section 3.1.1.8.3) [4].

References:

[1] Alex, R. F.; Pomeroy, R. K. (J. Organomet. Chem. **284** [1985] 379/84).
[2] Alex, R. F.; Pomeroy, R. K. (Organometallics **6** [1987] 2437/46).
[3] Alex, R. F.; Einstein, F. W. B.; Jones, R. H.; Pomeroy, R. K. (Inorg. Chem. **26** [1987] 3175/8).
[4] Arif, A. M.; Bright, T. A.; Heaton, D. E.; Jones, R. A.; Nunn, C. M. (Polyhedron **9** [1990] 1573/87).

3.1.1.8.2 Compounds with Seven Carbonyl Groups

The compounds dealt in Table 8 consist of $Os_3(CO)_7D_n$ or $Os_3(CO)_7(D\text{-}D)$ skeletons bearing additionally $\mu\text{-H}$ and/or $\mu_3\text{-S}$ or $\mu\text{-NO}$ ligands; D is a secondary or tertiary phosphane or $P(OCH_3)_3$ and D–D is $(C_6H_5)_2PCH_2P(C_6H_5)_2$ (abbreviated by dppm). The P-bonded ligands exhibited terminal, μ- or μ_3-bridging coordination modes, while the chelating D–D ligand is exclusively $\mu\text{-}\eta^2$ coordinated.

Table 8
Compounds with Seven Carbonyl Groups.
An asterisk preceding the compound number indicates further information at the end of the table, pp. 123/5.
Explanations, abbreviations, and units on p. X.

No. compound	method of preparation (yield) properties and remarks
1 Os$_3$(CO)$_7${P(OCH$_3$)$_3$}$_5$	by UV irradiation (Hanovia 200–W lamp) of Os$_3$(CO)$_8${P(OCH$_3$)$_3$}$_4$ (Section 3.1.1.8.3; in situ prepared) in heptane for 8 h in the presence of an excess of P(OCH$_3$)$_3$, followed by washing of the precipitate with hexane and toluene and purification by column chromatography on Florisil with CH$_2$Cl$_2$/P(OCH$_3$)$_3$ (6:1) as eluant (34%), along with small amounts of Os$_3$(CO)$_6${P(OCH$_3$)$_3$}$_6$ (Section 3.1.1.8.1) [2]

References on p. 125

Table 8 (continued)

No. compound	method of preparation (yield) properties and remarks

1 (continued)

by-product in the preparation of $Os_3(CO)_8\{P(OCH_3)_3\}_4$ from $Os_3(CO)_{12}$ and an excess of $P(OCH_3)_3$; see Section 3.1.1.8.3 [4]

orange, air-stable plates (from toluene or toluene/hexane at $-15\,°C$); m.p. 162.5 to 163 °C [2]

$^{13}C\ \{^1H\}$ NMR (toluene-d_8, $-44\,°C$): 200.7, 200.1 (br t, both axial CO of Os-2,3), 195.1 (d, axial CO of Os-1; J(P,C)=9.5, 2), 180.2 (s, equatorial CO of Os-1); spectra at $-44\,°C$ and $+75\,°C$ depicted [2]

$^{31}P\ \{^1H\}$ NMR (toluene-d_8, $-57\,°C$; vs. internal $P(OCH_3)_3$): -34.2 (P-1; $^3J(P,P)=6.4$), -36.0 (P-4; $^3J(P,P)=6.1$), -37.6 (P-2 or 5; $^3J(P,P)=7.6$), -38.0 (s, P-3), -42.8 (P-2 or 5; $^3J(P,P)=7.3$); at $+80\,°C$ only three singlets in the ratio 1:2:2 were observed; spectra between $-57\,°C$ and $+80\,°C$ depicted [2]

the variable-temperature $^{13}C\ \{^1H\}$ and $^{31}P\ \{^1H\}$ NMR determinations indicated that the P-1 ligand at Os-1 is probably involved in CO exchange processes by moving from one equatorial site to the other via a trigonal twist mechanism [2]

IR (CH_2Cl_2): 2033, 1952, 1902, 1889 (all CO) [2]

EI mass spectrum: $[M]^+$ [2]

UV irradiation in hexane in the presence of an excess of $P(OC_2H_5)_3$ led to $Os_3(CO)_6\{P(OCH_3)_3\}_5P(OC_2H_5)_3$ (Section 3.1.1.8.1) in low yields [4]

*2 $(\mu-H)Os_3(CO)_7\{\mu-P(C_6H_{11}-cyclo)_2\}_3$

from $Os_3(CO)_{12}$ and of $PH(C_6H_{11}-cyclo)_2$ (1:3) in refluxing toluene for 6 d, followed by extraction with toluene and precipitation by slow removal of the solvent and cooling to $-35\,°C$ (60%) [5]

orange, air-stable microcrystalline powder (from hexane at $-10\,°C$); m.p. 290 to 293 °C (dec.) [5]

^1H NMR (C_6D_6): 1.8 (br m, $C_6H_{11}-cyclo$), -17.18 (q, OsH, indicating a fluxional behavior for the hydride) [5]

$^{31}P\ \{^1H\}$ NMR (C_6D_6): 112.75 (s), indicating three equivalent phosphido bridges [5]

IR (Nujol mull): 2210, 1990, 1975, 1967, 1932 (all CO) [5]

References on p. 125

Table 8 (continued)

No.	compound	method of preparation properties and remarks

*3 $(\mu-H)_2Os_3(CO)_7(\mu_3-PC_4H_9-t)\{PH(C_4H_9-t)_2\}_2$

from $Os_3(CO)_{12}$ and $PH(C_4H_9-t)_2$ (1:3) in
refluxing n-butyl ether for 24 h, followed by
extraction with hexane and slow evaporation
of the solvent at $-35\,°C$ (68%) [5]
by thermolysis of $(\mu-H)_2Os_3(CO)_6$-
$\{\mu-P(C_4H_9-t)_2\}_2PH(C_4H_9-t)_2$ (Section
3.1.1.8.1), $(\mu-H)_2Os_3(CO)_8\{\mu-P(C_4H_9-t)_2\}_2$
(Section 3.1.1.8.3), $Os_3(CO)_9\{PH(C_4H_9-t)_2\}_3$
or $Os_3(CO)_{10}\{PH(C_4H_9-t)_2\}_2$ (both "Organoosmium Compounds" B 4b, in preparation)
in n-butyl ether for 24 h [5]
orange, air-stable crystals (from hexane at
$-10\,°C$); m.p. 158 to 160 °C [5]
^1H NMR (C_6D_6): 4.594 (d, PH; J(P,H)=342.5),
1.547 (d, CH_3 of $\mu_3-PC_4H_9-t$; J(P,H)=20.7),
0.916 (d, CH_3 of $PH(C_4H_9-t)_2$; J(P,H)=15.0),
-21.17 (t, OsH; ^{2}J(P,H)=10.2) [5]
^{31}P $\{^1H\}$ NMR (C_6D_6): 208.12 (d, μ_3-P;
^{2}J(P,P)=89.0), 14.08 (d, PH; ^{2}J(P,P)=89.1)
[5]
IR (Nujol mull): 2351, 2313 (both PH); 2066,
2043, 2008, 1989, 1948, 1921, 1913, 1261,
1170, 1066, 1026, 860, 851, 813, 675, 569, 528,
521 (no assignment given) [5]

4 $Os_3(CO)_7\{P(CH_3)_2C_6H_5\}_2(\mu_3-S)_2$

from $Os_3(CO)_9(\mu_3-S)_2$ ("Organoosmium Compounds" B 3, 1994, p. 196) and
$Pt\{P(CH_3)_2C_6H_5\}_4$ (ca. 1:1.5) in CH_2Cl_2 at
25 °C for 16 h under a CO atmosphere;
separation by TLC with CH_2Cl_2/hexane (1:4)
as eluant (32%), along with 4% of
$Os_3(CO)_8P(CH_3)_2C_6H_5(\mu_3-S)_2$ (Section
3.1.1.8.3) and 18% of $Os_3(CO)_9(\mu_3-S)_2$-
$Pt\{P(CH_3)_2C_6H_5\}_2$ ("Organoosmium Compounds" B 4b, in preparation) [3]

*5 $(\mu-H)_2Os_3(CO)_7(\mu-\eta^2-dppm)(\mu_3-S)$

from $(\mu-H)_2Os_3(CO)_9(\mu_3-S)$ ("Organoosmium
Compounds" B 3, 1994, p. 191) and
$(C_6H_5)_2PCH_2P(C_6H_5)_2$ (1:3) in refluxing
toluene for 16 h, followed by TLC with light
petroleum/CH_2Cl_2 (2:1) as eluant (34%);
similar preparation in refluxing octane for
4 h yielded the title compound in 15%, along
with 11% of $Os_3(CO)_8(\mu-\eta^2-dppm)_2$ (Section
3.1.1.8.3) [6]

References on p. 125

Table 8 (continued)

No. compound	method of preparation properties and remarks

*5 (continued)

pale yellow crystals (from hexane/CH₂Cl₂ or from hexane at −20 °C) [6]

^{1}H NMR (CDCl₃, 27 °C): 7.33 (m, C_6H_5), 4.66 (dtt, 1 H of CH_2), 3.56 (dt, 1 H of CH_2), −19.21 (dt, OsH-2; J=1.4, 8.8), −20.04 (dd, OsH-1; J=1.4, 32.5); for numbering, see Fig. 49; hydride region of spectra between −50 °C and +20 °C depicted [6]

variable-temperature ^{1}H NMR determinations between −50 °C and +20 °C indicated two separate exchange processes for μ-H-1 and μ-H-2 [6]

IR (CH₂Cl₂): 2057, 2032, 1989, 1976, 1956, 1927 (all CO) [6]

6   [(μ-H)₃Os₃(CO)₇(μ-η²-dppm)(μ₃-S)]PF₆

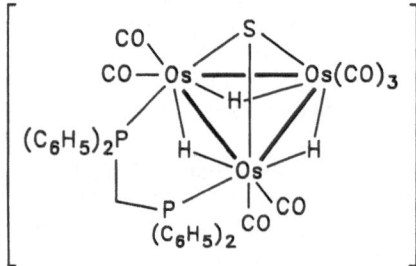

by protonation of No. 5 with CF₃CO₂H in CD₂Cl₂, followed by addition of NH₄PF₆ in CH₃OH and precipitation with small amounts of H₂O (80%) [6]

pale yellow crystals (from CH₂Cl₂/ether) [6]

^{1}H NMR (CD₂Cl₂): 7.7 to 7.1 (m, C_6H_5), 5.37 (dtt, 1 H of CH_2), 3.76 (dt, 1 H of CH_2), −19.04 (m, 1 OsH), −19.67 (m, 2 OsH); hydride region of the spectrum depicted; the spin system was analyzed as an ABB′MXX′ system; coupling is observed between the two ^{31}P nuclei, the three hydrides, and between the unique μ-H and one of the CH_2 protons [6]

^{1}H {^{31}P} NMR (CD₂Cl₂?): −19.10 (d, 1 OsH), −19.70 (dt, 2 OsH); the spectra are consistent with a static structure without hydride mobility revealing that each Os-Os edge is bridged by a μ-H ligand [6]

IR (CH₂Cl₂): 2118, 2057, 2045, 2001, 1975 (all CO) [6]

7   (μ-H)Os₃(CO)₇{P(OCH₃)₃}₃(μ-NO)

from (μ-H)Os₃(CO)₁₀(μ-NO) (Section 3.1.1.7.2.2) and P(OCH₃)₃ in refluxing cyclohexane at room temperature for 1.5 h, followed by TLC [1]

IR (cyclohexane): 2097, 2068, 2055, 2034, 2019, 2008, 1986, 1978, 1971, 1956 (all CO); (KBr): 1460 (NO) [1]

References on p. 125

*Further information:

(μ-H)Os$_3$(CO)$_7${μ-P(C$_6$H$_{11}$-cyclo)$_2$}$_3$ (Table **8**, No. **2**) crystallizes in the triclinic space group P1̄−C$_i^1$ (No. 2) with a = 13.042(2), b = 13.506(2), c = 13.807(2) Å, α = 99.64(1)°, β = 99.63(1)°, γ = 93.72(1)°; Z = 2, D$_c$ = 1.917 g/cm^3. The structure of the 48–electron molecule is shown in **Fig. 47**. Excluding metal–metal bonding, each Os atom has a roughly square–pyramidal coordination geometry. Os(1) and Os(2) may be considered to have formal oxidation states of (+1), while Os(3) would formally be (+2). Each bond edge of the triangular Os$_3$ core is bridged by a μ-P(C$_6$H$_{11}$-cyclo)$_2$ ligand; the μ-P(1) unit is bent below, while the μ-P(2) and μ-P(3) moieties are above the Os$_3$ plane. The hydride ligand although not located is assumed to bridge the Os(1)–Os(2) bond. The Os–Os and Os–P bonds fell within normal limits [5].

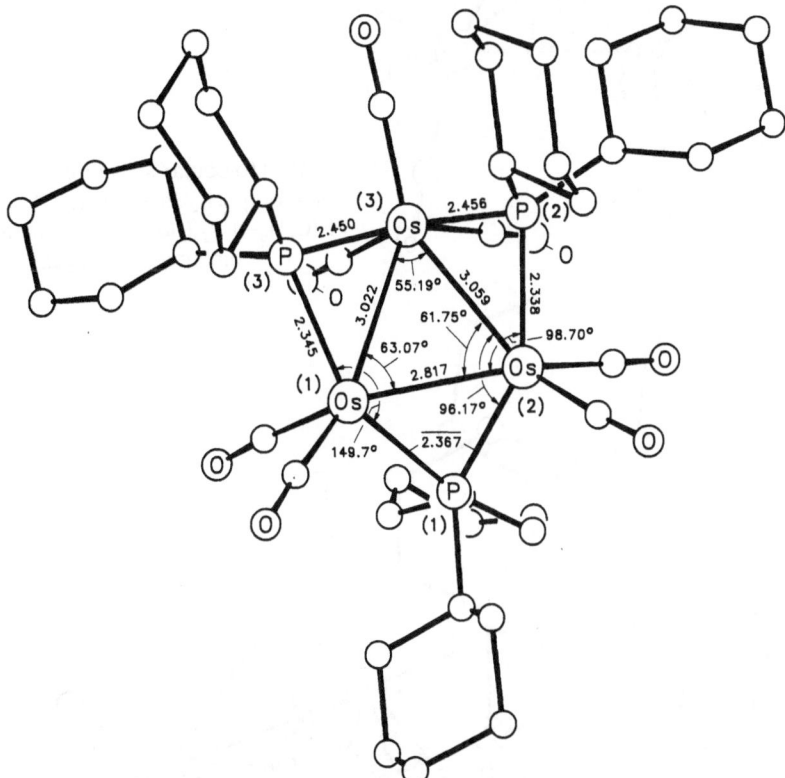

Fig. 47. Molecular structure of (μ-H)Os$_3$(CO)$_7${μ-P(C$_6$H$_{11}$-cyclo)$_2$}$_3$ (No. 2) with selected bond distances (in Å) and bond angles [5].

(μ-H)$_2$Os$_3$(CO)$_7$(μ$_3$-PC$_4$H$_9$-t){PH(C$_4$H$_9$-t)$_2$}$_2$ (Table **8**, No. **3**) crystallizes in the triclinic space group P1̄−C$_i^1$ (No. 2) with a = 12.211(3), b = 18.047(5), c = 18.412(5) Å, α = 96.97(2)°, β = 105.78(2)°, γ = 101.54(2)°; Z = 4, D$_c$ = 2.024 g/cm^3 with two independent molecules in the asymmetric unit cell. The molecules have similar geometric parameters. The structure of one molecule is shown in **Fig. 48**. The hydride ligands were not observed crystallographically, but they are assumed to bridge the Os(1)–Os(2) and Os(2)–Os(3) bond edges on the opposite side of the Os$_3$ triangle than the capping μ$_3$-PC$_4$H$_9$-t ligand. The Os(1)–μ$_3$-P

References on p. 125

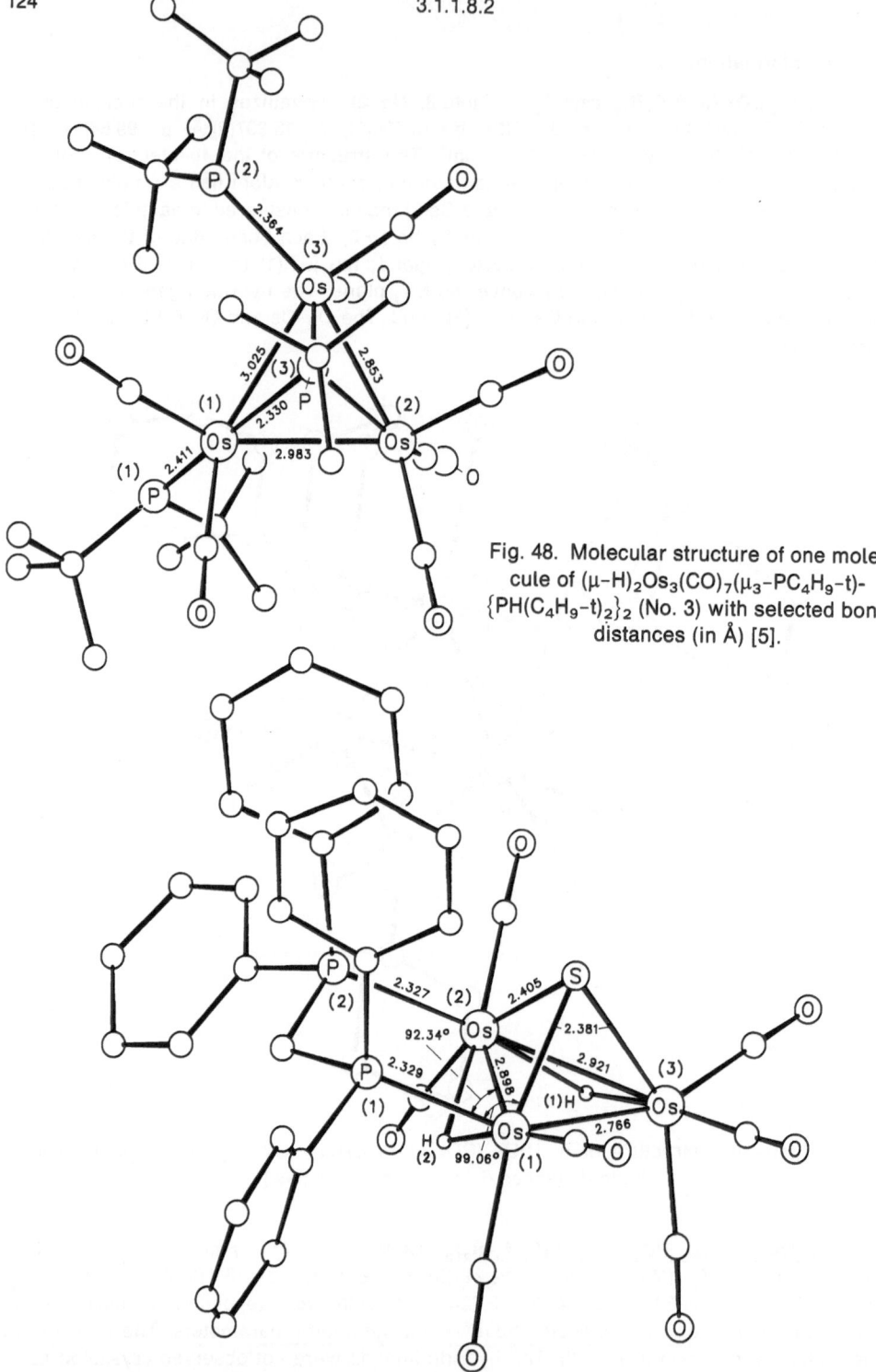

Fig. 48. Molecular structure of one mole-
cule of $(\mu\text{-H})_2Os_3(CO)_7(\mu_3\text{-PC}_4H_9\text{-t})$-
$\{PH(C_4H_9\text{-t})_2\}_2$ (No. 3) with selected bond
distances (in Å) [5].

Fig. 49. Molecular structure of $(\mu\text{-H})_2Os_3(CO)_7(\mu\text{-}\eta^2\text{-dppm})(\mu_3\text{-S})$ (No. 5) with selected bond
distances (in Å) and bond angles [6].

and the Os(3)-μ_3-P distances of ca. 2.330 Å are slightly longer than the $(CO)_3Os(2)-\mu_3$-P bond (2.306 Å). The Os-Os-Os angles range from 56.70° to 62.39° [5].

$(\mu\text{-}H)_2Os_3(CO)_7(\mu\text{-}\eta^2\text{-dppm})(\mu_3\text{-}S)$ (Table 8, No. 5) crystallizes in the monoclinic space group C2/c − C_{2h}^6 (No. 15) with a = 26.982(2), b = 13.202(1), c = 24.208(2) Å, β = 127.59(2)°; Z = 8, D_c = 2.30 g/cm³. The structure is shown in **Fig. 49**. The $\mu\text{-}\eta^2$-dppm ligand bridges the Os(1)-Os(2) bond edge with the P atoms occupying approximately equatorial positions although they are significantly above the Os_3 plane on the side of the face-capping sulfur atom. The hydrides were not located directly but their positions were calculated; they bridge the Os(1)-Os(2) and Os(2)-Os(3) bond edges on the opposite side of the Os_3 plane to the μ_3-S ligand [6].

Protonation with CF_3CO_2H and treatment with NH_4PF_6 led to $[(\mu\text{-}H)_3Os_3(CO)_7(\mu\text{-}\eta^2\text{-dppm})(\mu_3\text{-}S)]PF_6$ (No. 6) [6].

References:

[1] Johnson, B. F. G.; Raithby, P. R.; Zuccaro, C. (J. Chem. Soc. Dalton Trans. **1980** 99/104).
[2] Alex, R. F.; Pomeroy, R. K. (J. Organomet. Chem. **284** [1985] 379/84).
[3] Adams, R. D.; Horváth, I. T.; Wang, S. (Inorg. Chem. **25** [1986] 1617/23).
[4] Alex, R. F.; Pomeroy, R. K. (Organometallics **6** [1987] 2437/46).
[5] Arif, A. M.; Bright, T. A.; Heaton, D. E.; Jones, R. A.; Nunn, C. M. (Polyhedron **9** [1990] 1573/87).
[6] Azam, K. A.; Kabir, S. E.; Miah, A.; Day, M. W.; Hardcastle, K. I.; Rosenberg, E.; Deeming, A. J. (J. Organomet. Chem. **435** [1992] 157/67).

3.1.1.8.3 Compounds with Eight Carbonyl Groups

The compounds dealt with in this section generally consist of an $Os_3(CO)_8$ skeleton bearing P-bonded ligands such as phosphites or phosphanes (designated as D ligands) or chelating diphosphido ligands (designated as $\mu\text{-}\eta^2$-D-D ligands). Many compounds exhibit additional bridging or terminally coordinated hydrides and/or halogen ligands or O-, S-, and N-bonded moieties.

The D ligands of the type $P(OR)_3$ and PR_3 were found to be bonded terminally, occupying equatorial coordination sites, while PHR_2 and PH_2R also form $\mu\text{-}PR_2$ or μ_3-PR units, respectively.

In this section the D-D diphosphido ligands exclusively exhibit $\mu\text{-}\eta^2$ coordination. The chelating ligands were partially abbreviated by "dmpm", "dppm", "dppe", "dppp", and "dppen" for bis(dimethylphosphanyl)methane, bis(diphenylphosphanyl)methane, 1,2-bis(diphenylphosphanyl)ethane, 1,3-bis(dimethylphosphanyl)propane, and 1,1-bis(diphenylphosphanyl)ethene, respectively.

Most of the compounds listed in Table 9 were prepared by the following methods:

Method I:　Compounds of the type $(\mu\text{-}H)Os_3(CO)_8(\mu\text{-}\eta^2\text{-dppm})D_2$ were prepared from $(\mu\text{-}H)Os_3\{\mu_3\text{-}\eta^3\text{-}C_6H_4P(C_6H_5)CH_2P(C_6H_5)_2\}(CO)_8$ ("Organoosmium Compounds" B 6, 1993, p. 116) by treatment with one to four equivalents or an excess of D in toluene at room temperature for 1 min (in the case of phosphites) or 1 to 7 h (in the case of phosphanes), followed by TLC workup; D = $P(OCH_3)_3$, $P(OC_3H_7\text{-}i)_3$, $P(OC_4H_9\text{-}n)_3$, $P(OC_6H_5)_3$, $P(C_2H_5)_3$, $P(C_4H_9\text{-}n)_3$, $P(C_6H_5)_3$.
　　　　　　The products were generally contaminated with the initially formed $(\mu\text{-}H)Os_3\{\mu_3\text{-}\eta^3\text{-}C_6H_4P(C_6H_5)CH_2P(C_6H_5)_2\}(CO)_8D$ ("Organoosmium Compounds" B 5,

References on pp. 155/6

1994, pp. 319/20) and could not be completely separated in most cases. For $D = P(OCH_3)_3$, $P(OC_6H_5)_3$, and $P(C_6H_5)_3$, $Os_3(CO)_{10}(\mu-\eta^2\text{-dppm})$ ("Organo-osmium Compounds" B 4b, in preparation) was always obtained as a by-product with ca. 17% yield [41].

Method II: Compounds of the type $(\mu\text{-H})Os_3(CO)_8\{P(OCH_3)_3\}_2(\mu\text{-OR})$ were prepared from $(\mu\text{-H})Os_3(CO)_{10}(\mu\text{-OR})$ (R = H, CH_3, C_6H_5; see "Organoosmium Compounds" B 3, 1994, pp. 94, 96, and 102) and $P(OCH_3)_3$ in the presence of ca. 2.2 equivalents of $(CH_3)_3NO$ in CH_3CN at room temperature for ca. 20 min; workup by TLC with CH_2Cl_2/hexane (1:2) as eluant.

The reactions probably proceed via initially formed $(\mu\text{-H})Os_3(CO)_9(NCCH_3)$-$(\mu\text{-OR})$. For R = C_6H_5, a mixture of two geometrical isomers was obtained, which could be chromatographically separated [30].

Method III: Compounds of the type $[Os_3(CO)_8\{\mu-\eta^2\text{-(RO)}_2PN(C_2H_5)P(OR)_2\}_2(\mu\text{-X})]X$ were formed by reaction of $Os_3(CO)_8\{\mu-\eta^2\text{-(RO)}_2PN(C_2H_5)P(OR)_2\}_2$ (R = CH_3, C_3H_7-i, Nos. 32 or 33) with X_2 (X = Br or I) in a nonpolar solvent; the Cl-bridged derivatives were similarly produced under similar chlorinating conditions.

The products were identified and characterized, but obviously not isolated; no precise reaction conditions, yields, and spectroscopic data given [38].

Method IV: Compounds of the type $Os_3(CO)_8(\mu-\eta^2\text{-D-D})_2$ were prepared from $Os(CO)_{12}$ and D-D (ca. 1:2; D-D = dppm, dppen, 1,2-bis(diphenylphosphanyl)3,3,4,4-tetrafluorocyclobutene) in refluxing toluene or xylene for 8 to 9 h, followed by TLC with CH_2Cl_2/petroleum ether (1:2) as eluant [23, 24] or by column chromatography with ether/petroleum ether (3:1) as eluant [3].

Method V: Compounds of the type $(\mu\text{-H})_2Os_3(CO)_8(\mu-\eta^2\text{-D-D})$ (D-D = dppm, dppe, or dppp) were prepared by hydrogenation of $Os_3(CO)_{10}(\mu-\eta^2\text{-D-D})$ in refluxing toluene for 3 to 17 h by bubbling H_2 through the reaction mixture and purified by workup by TLC with CH_2Cl_2/hexane (1:9) or light petroleum/CH_2Cl_2 as eluants [34], or in toluene at 85 °C for 12 h under 25 atm of H_2 in an autoclave, followed by removal of the solvent and recrystallization [22].

Method VI: Compounds of the type $(\mu\text{-H})Os_3(CO)_8(\mu-\eta^2\text{-D-D})(\mu\text{-SR})$ were prepared from $Os_3(CO)_9(\mu-\eta^2\text{-D-D})NCCH_3$ (D-D = dmpm, dppm; "Organoosmium Compounds" B 4b, in preparation) and an excess of RSH (R = CH_3, C_2H_5, C_4H_9-t, C_6H_5) in refluxing CH_2Cl_2 for 5 h [31] or at room temperature for 40 h [39]; workup by TLC with ethyl acetate/hexane (1:5) [31] or petroleum ether/CH_2Cl_2 (10:3) [39] as eluants. $Os_3(CO)_{10}(\mu-\eta^2\text{-dppm})$ ("Organoosmium Compounds" B 4b, in preparation) was a by-product [31].

$Os_3(CO)_8\{P(C_6H_5)_2\}P(C_6H_5)_3$ was obtained in a mixture with eight other products by reaction of $Os_3(CO)_{12}$ with 2 equivalents of $P(C_6H_5)_3$. Probably both P ligands are exclusively bonded to the Os_3 core by the P atoms in contrast to compounds where deprotonated $P(C_6H_5)_3$ moieties are additionally coordinated by one or two Os-C bonds ("Organoosmium Compounds" B 5, 1994, p. 261, B 6, 1993, p. 93, and B 7, in preparation). However, structural details are not known, because crystals suitable for X-ray analysis could not be obtained [4].

$Os_3(CO)_8\{PH(C_6H_5)_2\}_4$ (purple oil at room temperature) was allegedly formed in the reaction of $(\mu\text{-H})_3Os_3(CO)_9NiC_5H_5$ with $PH(C_6H_5)_2$ in the presence of $(CH_3)_3NO$, based on ^1H NMR spectra and elemental analysis indicating the absence of a C_5H_5Ni fragment. However, this formulation appears to be inconsistent with the simplicity of ^{31}P NMR and IR spectra.

^{1}H NMR (CDCl$_3$): 8.89, 7.11 (s's, PH; ^{1}J(P,H)=480 Hz), 7.74 to 7.40 (m, C$_6$H$_5$) ppm. ^{31}P NMR (CDCl$_3$): 19.67 (s) ppm. IR (hexane): 2065, 2024, 2008 (all CO) cm^{-1}; a spectrum in hexane/CHCl$_3$ is depicted [25].

Os$_3$(CO)$_8${P(CH$_3$)$_2$C$_6$H$_5${$_4$ was probably obtained as a by-product in the preparation of (μ-H)$_3$Os$_3$(CO)$_7${P(CH$_3$)$_2$C$_6$H$_5${$_2$NiC$_5$H$_5$ and (μ-H)$_3$Os$_3$(CO)$_8$P(CH$_3$)$_2$C$_6$H$_5$NiC$_5$H$_5$ ("Organoosmium Compounds" B 4b, in preparation) by treatment of (μ-H)$_3$Os$_3$(CO)$_9$NiC$_5$H$_5$ with P(CH$_3$)$_2$C$_6$H$_5$ [25].

In Table 9 the compounds are arranged according to the type of the P ligand: compounds with terminal monodentate D ligands such as phosphanes or phosphites, compounds with μ-PR$_2$ and μ$_3$-PR ligands, a compound with an η^2 ligand bonded by one P and one N atom, and compounds with bridging bisphosphane ligands of the η^2-D-D type. Within these groups the compounds are arranged by the nature of additional ligands.

Table 9
Compounds with Eight Carbonyl Groups.
An asterisk preceding the compound number indicates further information at the end of the table, pp. 148/55.
Explanations, abbreviations, and units on p. X.

No. compound	method of preparation (yield) properties and remarks

compound of the type Os$_3$(CO)$_8$D$_4$

1 Os$_3$(CO)$_8${P(OCH$_3$)$_3${$_4$

Isomer 1

Isomer 2

rapid, nondissociative equilibrium mixture of Isomers 1 and 2 (ca. 1.3:1 in toluene at −86 °C), based on ^{31}P {^{1}H} NMR spectra; the existence of a third isomer cannot be completely ruled out [26]
from Os$_3$(CO)$_{12}$ and an excess of P(OCH$_3$)$_3$ in heptane at 140 to 145 °C for 24 h [17, 26, 27], followed by chromatography with toluene or toluene/CH$_2$Cl$_2$ as eluant (no yield given), along with small amounts of Os$_3$(CO)$_7${P(OCH$_3$)$_3${$_5$ (Section 3.1.1.8.2); only small amounts of the title compound in xylene at 125 °C for 20 h [26]
by-product in the preparation of Os$_3$(CO)$_{12-n}${P(OCH$_3$)$_3${$_n$ (n=1, 2, or 3; see "Organoosmium Compounds" B 4b, in preparation) from Os$_3$(CO)$_{12}$ and an excess of P(OCH$_3$)$_3$ in refluxing toluene for 16 h [33]
pale orange, air-stable plates (from hexane/toluene or hexane/CH$_2$Cl$_2$); m.p. 163 to 165 °C (dec.) [26]
^{1}H NMR (CDCl$_3$): 3.62 (d, CH$_3$; J(P,H)=12.2), 3.61 (d, CH$_3$; J(P,H)=11.8) [26]
^{13}C NMR (toluene-d$_8$, −83 °C): 200.2 (br t, 2 CO, CO-a; J(P,C)=10.8), 198.8 (br t, 2 CO,

References on pp. 155/6

Table 9 (continued)

No. compound	method of preparation (yield) properties and remarks

1 (continued)

CO-x; J(P,C) = 10), 195.1 (br s, 6 CO, CO-y and CO-b or CO-d), 194.1 (d, 2 CO, CO-b or CO-d; J(P,C) = 6.4), 181.9 (s, CO-c or CO-e), 180.8 (s, 2 CO, CO-z), 178.7 (s, 1 CO, CO-c or CO-e); at $-70\,°C$ and 25.18 MHz the broad triplets 200.0 and 198.8 ppm were well resolved; (toluene-d_8, + 80 °C, 100.6 MHz): 198.4 (t; J(P,C) = 11.2), 189.1 (s); various spectra between $-83\,°C$ and $+80\,°C$ depicted [26]

^{31}P {^1H} NMR (toluene/toluene-d_8, $-81\,°C$; vs. internal $P(OCH_3)_3$): -33.4 (d; J(P,P) = 4.4), -33.6 (s), -35.0(s), -36.4 (d; J(P,P) = 4.2), -36.7 (br s), -39.6 (d; J(P,P) = 3.9); (toluene/toluene-d_8 (5:1), $+53\,°C$): -38.5 (s, intensity = 1), -40.1 (s, intensity = 1); no assignment given; various spectra between $-76\,°C$ and $+61\,°C$ depicted [26]

variable-temperature ^{13}C {^1H} and ^{31}P {^1H} NMR between $-80\,°C$ and $+80\,°C$ revealed separate exchange processes: a lower-energy process involves the six axial CO groups rotating in a plane perpendicular to the Os_3 plane by a pairwise terminal-bridge exchange; a higher-energy process involves the $P(OCH_3)_3$ ligands of an $Os_3(CO)_3P(OCH_3)_3$ unit moving from one equatorial site to the other via a restricted trigonal-twist mechanism; this simultaneously involves axial-equatorial carbonyl exchange; the $P(OCH_3)_3$ ligand never enters an axial position probably for steric reasons [26]

IR (CH_2Cl_2): 2049, 1989, 1969, 1941, 1909 (all CO) [26]

EI mass spectrum: [M]$^+$, [M$-$n CO]$^+$, n = 1 to 5 [26]

UV irradiation (Hanovia 200-W lamp) in heptane for 48 h in the presence of an excess of $P(OCH_3)_3$ yielded $Os_3(CO)_6${$P(OCH_3)_3$}$_6$ (Section 3.1.1.8.1) [27]; similar UV irradiation for only 8 h yielded mainly $Os_3(CO)_7${$P(OCH_3)_3$}$_5$ (Section 3.1.1.8.2) and only minor amounts of $Os_3(CO)_6${$P(OCH_3)_3$}$_6$ (Section 3.1.1.8.1) [17]

Table 9 (continued)

No. compound	method of preparation (yield) properties and remarks

compounds of the types (μ-H)Os₃(CO)₈D₂(μ-X) and Os₃(CO)₈D₂(μ-X)₂ (X = halogen)

2 (μ-H)Os₃(CO)₈{P(C₆H₅)₃}₂(μ-Cl)

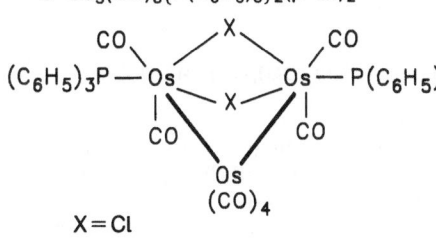

from (μ-H)Os₃(CO)₁₀(μ-Cl) ("Organoosmium
    Compounds" B 3, 1994, p. 82) and P(C₆H₅)₃
    (1:2) in CH₂Cl₂/CH₃CN at room temperature
    for 15 min in the presence of 2.5 equivalents
    of (CH₃)₃NO; purified by chromatography
    with CH₂Cl₂/hexane (3:1) as eluant (59%),
    along with (μ-H)Os₃(CO)₉P(C₆H₅)₃(μ-Cl) (see
    "Organoosmium Compounds" B 4b, in
    preparation) [28]
by-product in the preparation of
    (μ-H)Os₃(CO)₉P(C₆H₅)₃(μ-Cl) from
    (μ-H)Os₃(CO)₉N(CH₃)₃(μ-Cl) and P(C₆H₅)₃ in
    CH₂Cl₂ at room temperature for 2 to 3 h;
    workup as before (10%) [28]
¹H NMR (CD₂Cl₂): 7.24 (m, C₆H₅), −12.71 (t,
    OsH; J(P,H)=6.7) [28]
IR (cyclohexane): 2079, 2015, 2008, 2002, 1958,
    1947 (all CO) [28]

3 Os₃(CO)₈{P(C₆H₅)₃}₂(μ-Cl)₂

CO   X   CO
(C₆H₅)₃P —Os   Os— P(C₆H₅)₃
    | X |
    CO   CO
              Os
             (CO)₄
X = Cl

from Os₃(CO)₁₀(μ-Cl)₂ ("Organoosmium Com-
    pounds" B 3, 1994, p. 71) and P(C₆H₅)₃ in
    refluxing benzene for 2 h [1]
yellow plates (from pentane); m.p. 254 to
    255 °C (dec.) [1]
IR (cyclohexane): 2079, 2014, 2004, 1967, 1943
    (all CO), 289, 266, 258 (all OsCl) [1]
mass spectrum: [M]⁺, [M−P(C₆H₅)₃]⁺ [1]
treatment with Cl₂ in CHCl₃ for 15 s led to
    cis-Os(CO)₄Cl₂ ("Organoosmium Com-
    pounds" A 1, 1993, p. 204) and
    Os₂(CO)₄{P(C₆H₅)₃}₂Cl₄ [1]

4 Os₃(CO)₈{P(C₆H₅)₃}₂(μ-I)₂
    as No. 3, but X = I

from Os₃(CO)₁₀(μ-I)₂ ("Organoosmium Com-
    pounds" B 3, 1994, p. 72) and P(C₆H₅)₃ (1:3)
    in refluxing cyclohexane for 2.5 h, followed
    by filtration of the precipitate and recrystalli-
    zation (97%) [12]
light yellow crystals (from n-hexane); m.p. 218
    to 220 °C (dec.) [12]
¹H NMR (acetone-d₆): 7.44 (C₆H₅) [12]
IR (CHCl₃): 2077, 2009, 1963, 1940 (all CO) [12]
polarography in CH₃CN (10⁻³ M, 0.1 M
    [N(C₂H₅)₄]BF₄, dropping Hg electrode)
    showed two irreversible one-electron reduc-
    tions at −1.79 and −2.61 V vs. Ag/Ag⁺ [12]

References on pp. 155/6

Table 9 (continued)

No. compound	method of preparation (yield) properties and remarks
5 $Os_3(CO)_8\{P(OC_6H_5)_3\}_2(\mu\text{-Br})_2$	from $Os_3(CO)_{10}(\mu\text{-Br})_2$ ("Organoosmium Compounds" B 3, 1994, p. 72) and $P(OC_6H_5)_3$ (ca. 1:3) in CH_3CN at 70 °C for 6 h, followed by column chromatography with benzene as eluant (95%) [12] yellow crystals; m.p. 168 to 170 °C (dec.) [12] ^1H NMR ($CDCl_3$): 7.14 (C_6H_5) [12] IR ($CHCl_3$): 2082, 2026, 2003, 1957 (all CO) [12] polarography in CH_3CN (10^{-3} M, 0.1 M $[N(C_2H_5)_4]BF_4$, dropping Hg electrode) showed two irreversible one-electron reductions at -1.70 and -2.39 V vs. Ag/Ag$^+$ [12]
6 $Os_3(CO)_8\{(C_6H_5)_2P\text{-Pol}\}_2(\mu\text{-Cl})_2$ (Pol = Block copolymer from styrene and divinylbenzene)	from $Os_3(CO)_{10}(\mu\text{-Cl})_2$ ("Organoosmium Compounds" B 3, 1994, p. 71) and phoshane-functionalized poly(styrene-divinylbenzene) in THF (no details given) [16] IR (no medium given): 2096, 2073, 2036, 2007, 1997, 1959. 1939 (all CO); spectrum depicted [1 relatively stable; showed only low catalytic activity for 1-butene isomerization below 125 °C and for ethene hydrogenation below 160 °C [16]

compounds of the types $(\mu\text{-H})Os_3(CO)_8D_2(\mu\text{-OR})$ and $(\mu\text{-H})Os_3(CO)_8D_2\{\mu\text{-}\eta^2\text{-}O_2CR\}$

7 $(\mu\text{-H})Os_3(CO)_8\{P(OCH_3)_3\}_2(\mu\text{-OH})$ 　　II (59%) [30]

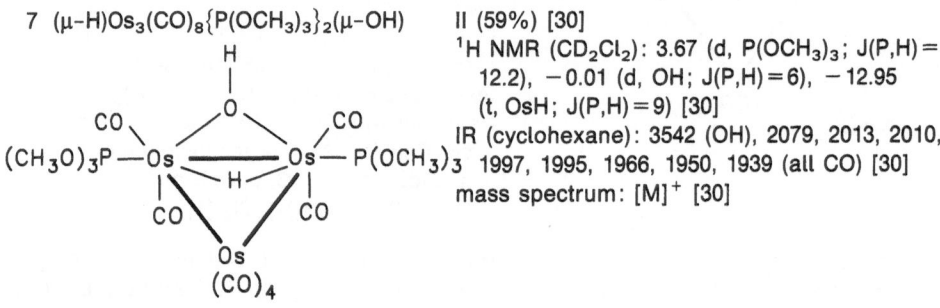

^1H NMR (CD_2Cl_2): 3.67 (d, $P(OCH_3)_3$; $J(P,H) = 12.2$), -0.01 (d, OH; $J(P,H) = 6$), -12.95 (t, OsH; $J(P,H) = 9$) [30]
IR (cyclohexane): 3542 (OH), 2079, 2013, 2010, 1997, 1995, 1966, 1950, 1939 (all CO) [30]
mass spectrum: $[M]^+$ [30]

8 $(\mu\text{-H})Os_3(CO)_8\{P(OCH_3)_3\}_2(\mu\text{-OCH}_3)$ 　　II (50%) [30]

^1H NMR (CD_2Cl_2): 3.79 (probably $\mu\text{-OCH}_3$), 3.78, 3.47 (d's, each 9 H, both $P(OCH_3)_3$; $J(P,H) = 12.05$), -12.82 (dd, OsH; $J(P,H) = 8$, 65) [30]; no assignment for reported resonances at 7.04 (m, 2 H) and 6.65 (t, 1 H; $J(H,H) = 7.2$) given [30]
IR (cyclohexane): 2075, 2011, 1993, 1990, 1963, 1956, 1943, 1940 (all CO) [30]
mass spectrum: $[M]^+$ [30]

References on pp. 155/6

Table 9 (continued)

No. compound	method of preparation (yield) properties and remarks

9 (μ-H)Os$_3$(CO)$_8${P(OCH$_3$)$_3$}$_2$(μ-OC$_6$H$_5$)

C$_6$H$_5$
|
CO   O   P(OCH$_3$)$_3$
\   / \   /
(CH$_3$O)$_3$P — Os ═══════ Os — CO
| \ H / |
CO   \   /   CO
Os
(CO)$_4$

isomer of No. 10, differing in the position of a
P(OCH$_3$)$_3$ moiety, forming a 3:2 equilibrium
with No. 10 in CH$_2$Cl$_2$ [30]
II (30%) [30]
^{1}H NMR showed a dd for OsH (no values
given) [30]
IR (cyclohexane): 2077, 2014, 1997, 1992, 1966,
1950, 1947 (all CO) [30]
mass spectrum: [M]$^+$ [30]

10 (μ-H)Os$_3$(CO)$_8${P(OCH$_3$)$_3$}$_2$(μ-OC$_6$H$_5$)

C$_6$H$_5$
|
CO   O   CO
\   / \   /
(CH$_3$O)$_3$P — Os ═══════ Os — P(OCH$_3$)$_3$
| \ H / |
CO   \   /   CO
Os
(CO)$_4$

isomer of No. 9; see there
II (25%) [30]
^{1}H NMR (CD$_2$Cl$_2$): 3.63 (d, P(OCH$_3$)$_3$; J(P,H)=
12.0), −11.60 (t, OsH; J(P,H)=8.6) [30];
values and assignments for C$_6$H$_5$, given in
[30], seem to be dubious
IR (cyclohexane): 2080, 2016, 1999, 1968, 1953
(all CO) [30]
mass spectrum: [M]$^+$ [30]

11 (μ-H)Os$_3$(CO)$_8${P(C$_6$H$_5$)$_3$}$_2${μ-η^2-O$_2$CC$_5$H$_4$Mn(CO)$_3$}

from (μ-H)Os$_3$(CO)$_{10}${μ-η^2-O$_2$CC$_5$H$_4$Mn(CO)$_3$}
("Organoosmium Compounds" B 3, 1994,
p. 128) and P(C$_6$H$_5$)$_3$ in THF at room temperature for 3 h in presence of (CH$_3$)$_3$NO (55%);
no further data given [19]

compounds of the type (μ-H)Os$_3$(CO)$_8$D$_2$(μ-SR) and ions derived therefrom

12 (μ-H)Os$_3$(CO)$_8${P(C$_2$H$_5$)$_3$}$_2$(μ-SC$_6$H$_5$)

C$_6$H$_5$
|
(C$_2$H$_5$)$_3$P   S   P(C$_2$H$_5$)$_3$
\   / \   /
CO — Os ═══════ Os — CO
| \ H / |
CO   \   /   CO
Os
(CO)$_4$

isomer of No. 13, differing in positions of the
P(C$_2$H$_5$)$_3$ ligands, based on ^{1}H NMR spectra
[2]
from (μ-H)Os$_3$(CO)$_{10}$(μ-SC$_6$H$_5$) ("Organoosmium Compounds" B 3, 1994, p. 162) and
an excess of P(C$_2$H$_5$)$_3$ in benzene on reflux
for 3 h; purified by column chromatography
on alumina with toluene/pentane (1:1) as
eluant; separation from No. 13 by fractional
recrystallization from pentane;
(μ-H)Os$_3$(CO)$_9$P(C$_2$H$_5$)$_3$(μ-SC$_6$H$_5$) (see
"Organoosmium Compounds" B 4b, in
preparation) was obtained as a co-product
[2]
yellow needles (from pentane); m.p. 128 to
130 °C [2]

References on pp. 155/6

Table 9 (continued)

No. compound	method of preparation (yield) properties and remarks
12 (continued)	^1H NMR $(CDCl_3)$: -16.46 (t, OsH; $J(P,H)=30.2$) [2] IR (cyclohexane): 2067, 2063, 1998, 1990, 1958, 1949, 1937, 1930 (all CO) [2] protonation of Nos. 12 and 13 with concentrated H_2SO_4 yielded No. 15 [2]

13 $(\mu-H)Os_3(CO)_8\{P(C_2H_5)_3\}_2(\mu-SC_6H_5)$

isomer of No. 12; for preparation and protonation see there
orange-red prisms (from pentane); m.p. 142 to 144 °C [2]
^1H NMR $(CDCl_3)$: -16.66 (d, OsH; $J(P,H)=26.0$) [2]
IR (cyclohexane): 2067, 2058, 2015, 1986, 1982, 1950, 1943, 1929 [2]

14 $[(\mu-H)_2Os_3(CO)_8\{P(C_2H_5)_3\}_2(\mu-SC_6H_5)]PF_6$

mixture of two isomers α and β which probably differ only in the C_6H_5 arrangement [2]
from No. 15 by treatment with aqueous NH_4PF_6 [2]
pale yellow powder [2]
^1H NMR (CH_3NO_2):
Isomer α: -18.69 (q, 1 OsH; $J(P,H)=8.1$, 11.5), -15.73 (d, 1 OsH; $J(P,H)=25.5$);
Isomer β: -19.37 (q, 1 OsH; $J(P,H)=12.5$, 16.7), -15.75 (d, 1 OsH; $J(P,H)=22.0$) [2]
the structure assumedly corresponds to that of No. 13 having one $P(C_2H_5)_3$ ligand at the unique Os center and the other one to the Os atom which is also coordinated to the SC_6H_5 ligand [2]
IR (CH_3NO_2): 2130, 2104, 2096, 2062, 2051, 2014, 2001, 1985 (all CO) [2]

15 $[(\mu-H)_3Os_3(CO)_8\{P(C_2H_5)_3\}_2(\mu-SC_6H_5)]SO_4$

by protonation of Nos. 12 and 13 with concentrated H_2SO_4 (not isolated) [2]
^1H NMR $(H_2SO_4$, vs. internal dimethyl sulfoxide): -14.94 (d, 1 OsH; $J(P,H)=16.0$), -19.44 (q, 1 OsH; $J(P,H)=18.0$, 10.2), -20.73 (d, 1 OsH; $J(P,H)=11.2$); two of the hydrido ligands are coupled to one ^{31}P nucleus, while the third is coupled to both of the ^{31}P nuclei [2]

References on pp. 155/6

Table 9 (continued)

No. compound	method of preparation (yield) properties and remarks

treatment with aqueous NH_4PF_6 yielded No. 14 by partial deprotonation [2]

compounds of the type $Os_3(CO)_8D(\mu_3-S)_2$

16 $Os_3(CO)_8PH(C_6H_5)_2(\mu_3-S)_2$

from $Os_3(CO)_9(\mu_3-S)_2$ ("Organoosmium Compounds" B 3, 1994, p. 196) and $PH(C_6H_5)_2$ (1:1) in hexane at room temperature for 3 h; purified by TLC with hexane/CH_2Cl_2 (4:1) as eluant (93%) [13]

yellow crystals (from hexane) [13]

^1H NMR ($CDCl_3$): 7.73 (d, PH; J(P,H)=397.9), 7.73 to 7.30 (m, C_6H_5) [13]

IR (hexane): 2084, 2065, 2052, 2024, 2009, 1991, 1983, 1965 (all CO) [13]

decarbonylation in refluxing octane for 20 h gave $Os_6(CO)_{14}\{\mu-P(C_6H_5)_2\}(\mu_4-S)(\mu_3-S)_3$ ("Organoosmium Compounds" B 9, 1995, p. 159) [13]

*17 $Os_3(CO)_8P(CH_3)_2C_6H_5(\mu_3-S)_2$

from $Os_3(CO)_9(\mu_3-S)_2$ ("Organoosmium Compounds" B 3, 1994, p. 196) and $P(CH_3)_2C_6H_5$ in hexane at room temperature for 3 h; workup by TLC with CH_2Cl_2/hexane (5:95) as eluant (88%) [11]

from $Os_3(CO)_9(\mu_3-S)_2$ and $Pt\{P(CH_3)_2C_6H_5\}_4$ (ca. 1:1.5) in CH_2Cl_2 at 25 °C for 16 h under a CO atmosphere; separation by TLC with CH_2Cl_2/hexane (1:4) as eluant (4%), along with 32% of $Os_3(CO)_7\{P(CH_3)_2C_6H_5\}_2(\mu_3-S)_2$ (Section 3.1.1.8.2) and $Os_3(CO)_9(\mu_3-S)_2Pt-\{P(CH_3)_2C_6H_5\}_2$ [21]

by carbonylation of $Os_3(CO)_9(\mu_3-S)_2-W(CO)_3P(CH_3)_2C_6H_5$ in CH_2Cl_2 at room temperature for 18 h under 1 atm of CO; workup by TLC (10%), along with $Os_3(CO)_9(\mu_3-S)_2$, $W(CO)_5P(CH_3)_2C_6H_5$, and $Os_3(CO)_{10}(\mu_3-S)_2-W(CO)_2\{P(CH_3)_2C_6H_5\}_2$ [15]

by-product in the preparation of $Os_3(CO)_8(\mu_4-S)(\mu_3-S)W_2(CO)_6\{P(CH_3)_2C_6H_5\}_2$ by photolysis of $Os_3(CO)_9(\mu_3-S)_2-W(CO)_3P(CH_3)_2C_6H_5$ and $W(CO)_5P(CH_3)_2C_6H_5$ in hexane for 2 h; workup by TLC with CH_2Cl_2/hexane (3:7) as eluant (trace amounts), along with $Os_3(CO)_9(\mu_3-S)_2$ [15]

References on pp. 155/6

Table 9 (continued)

No. compound	method of preparation (yield) properties and remarks

*17 (continued)

by photolysis of $Os_3(CO)_{10}(\mu_3-S)_2W(CO)_2$-$\{P(CH_3)_2C_6H_5\}_2$ in hexane for 1 h; workup as before (trace amounts), along with $Os_3(CO)_9(\mu_3-S)_2$, $W(CO)_5P(CH_3)_2C_6H_5$, 18% of $Os_3(CO)_9(\mu_3-S)_2W(CO)_3P(CH_3)_2C_6H_5$, and 37% of $Os_3(CO)_9(\mu_3-S)_2W(CO)_2$-$\{P(CH_3)_2C_6H_5\}_2$ [15]

yellow, air–stable crystals (from hexane) [11]

^{1}H NMR (CDCl$_3$): 7.72 to 7.40 (m, C$_6$H$_5$); 2.51 (d, CH$_3$; J(P,H) = 11.0) [11]

IR (hexane): 2083, 2065, 2050, 2018, 2008, 1995, 1974, 1962 (all CO) [11]

18 $Os_3(CO)_8P(C_6H_5)_3(\mu_3-S)_2$

from $Os_3(CO)_9(\mu_3-S)_2$ ("Organoosmium Compounds" B 3, 1994, p. 196) and $Pt(C_2H_4)\{P(C_6H_5)_3\}_2$ (ca. 1:1) in THF at 0 °C for 30 min and room temperature for 2 h; purification by TLC with hexane/CH$_2$Cl$_2$ (ca. 2:1) as eluant (11%), along with 34% of $Os_3(CO)_{10}P(C_6H_5)_3(\mu_3-S)_2PtP(C_6H_5)_3$ and 24% of $Os_3(CO)_9(\mu_3-S)_2PtP(C_6H_5)_3$ (both "Organoosmium Compounds" B 4b, in preparation); similar preparation under a CO atmosphere gave the same yield, but 76% of $Os_3(CO)_{10}(\mu_3-S)_2PtP(C_6H_5)_3$ and only 3% of $Os_3(CO)_9P(C_6H_5)_3(\mu_3-S)_2PtP(C_6H_5)_3$ [14]

by thermolysis of $Os_3(CO)_{10}(\mu_3-S)_2PtP(C_6H_5)_3$ in refluxing toluene for 4.5 h under an atmosphere of CO; workup by TLC (31%), along with $Os_3(CO)_9(\mu_3-S)_2$ [14]

from $Os_3(CO)_{10}(\mu_3-S)_2PtP(C_6H_5)_3$ and $P(C_6H_5)_3$ in refluxing hexane for 3 h; workup by TLC with CH$_2$Cl$_2$/hexane (1:5) as eluant (17%), along with 40% of $Os_3(CO)_9P(C_6H_5)_3(\mu_3-S)_2$-$PtP(C_6H_5)_3$ as the main product [14]

IR (hexane): 2082, 2051, 2023, 2007, 1989, 1981, 1965 (all CO) [14]

an X–ray analysis revealed the monoclinic space group $P2_1/c - C_{2h}^5$ (No. 14) with a = 10.459(5), b = 9.705(7), c = 29.20(3) Å, β = 94.50(5)°; no further data or figure given [14]

compounds with ligands of the D type and additional N–bonded ligands

*19 $Os_3(CO)_8P(OCH_3)_3(NO)_2$

from $Os_3(CO)_9(NO)_2$ ("Organoosmium Compounds" B 3, 1994, p. 247) and $P(OCH_3)_3$ in cyclohexane at 25 °C [5] or in n–heptane at

Table 9 (continued)

No. compound	method of preparation (yield) properties and remarks
	room temperature for 3 h; workup by TLC with CHCl$_3$/cyclohexane (2:3) as eluant (68%) [7]
	red elongated plates (from hexane) [5, 7]
	^{13}C NMR (no medium given, 30 °C): 186.1, 184.2 (each 2 axial CO), 176.6, 174.1, 173.8, 172.4 (each 1 equatorial CO) [7]
	IR (cyclohexane): 2104, 2056, 2021, 2006, 1993, 1985, 1979 (all CO), 1705, 1672 (both NO) [5, 7]
	mass spectrum: [M]$^+$ [7]
20 Os$_3$(CO)$_8$P(C$_6$H$_5$)$_3$(NO)$_2$	from Os$_3$(CO)$_9$(NO)$_2$ ("Organoosmium Compounds" B 3, 1994, p. 247) and an excess of P(C$_6$H$_5$)$_3$ as described for No. 19 (80%) [7]
	red crystals (from hexane) [7]
	IR (cyclohexane): 2100, 2052, 2020, 2003, 1988, 1971 (all CO), 1700, 1671 (both NO) [7]
	mass spectrum: [M]$^+$ [7]
	carbonylation in refluxing n-heptane for 30 min led to Os$_3$(CO)$_9$P(C$_6$H$_5$)$_3$(μ-NO)$_2$ (see "Organoosmium Compounds" B 4b, in preparation) [7]

21 (μ-H)Os$_3$(CO)$_8${P(OCH$_3$)$_3$}$_2$(μ-NHCH$_3$)

Isomer 1

Isomer 2

ca. 1:1 equilibrium mixture of two isomers; the proposed structure for Isomers 1 and 2 is reproduced in the bystanding figures [29]

from (μ-H)Os$_3$(CO)$_9$P(OCH$_3$)$_3$(μ-NHCH$_3$) (Isomer 3, "Organoosmium Compounds" B 4b, in preparation) and P(OCH$_3$)$_3$ in refluxing cyclohexane; workup by TLC with ethyl acetate/hexane (1:4) as eluant (75%) [29]

^{1}H NMR (CDCl$_3$):
Isomer 1: 3.98 (br, NH), 3.75, 3.59 (d's, both POCH$_3$; J(P,H) = 11.4, 12.3), 3.24 (d, NCH$_3$; J(H,H) = 5.9), −14.94 (dt, OsH; J(H,H) = 2.2, J(P,H) = 2.2, J(P,H) = 56.5);
Isomer 2: 4.07 (br, NH), 3.68, 3.62 (d, both POCH$_3$; J(P,H) = 11.9, 12.3), 3.03 (d, NCH$_3$; J(H,H) = 6.3), −14.87 (m, OsH) [29]

IR (cyclohexane):
Isomer 1: 2068, 2023, 1995, 1965, 1962, 1953, 1943 (all CO);
Isomer 2: 2068, 2023, 1995, 1990, 1968, 1953, 1943, 1938 (all CO) [29]

mass spectrum: [M]$^+$ [29]

References on pp. 155/6

Table 9 (continued)

No. compound	method of preparation (yield) properties and remarks

22 $(\mu-H)Os_3(CO)_8P(CH_3)_2C_6H_5\{\mu_3-\eta^2-N(C_3H_7-i)CHNC_3H_7-i\}$

by thermolysis of $(\mu-H)Os_3(CO)_9P(CH_3)_2C_6H_5$-$\{\mu-\eta^2-N(C_3H_7-i)CH=NC_3H_7-i\}$ (see "Organo-osmium Compounds" B 4b, in preparation) in refluxing octane for 3 h, followed by chromatography; only obtained as an inseparable 1:1 mixture with $(\mu-H)Os_3(CO)_9\{\mu_3-\eta^2-N(C_3H_7-i)CH=NC_3H_7-i\}$ (Section 3.1.1.7.3.3) [9]

IR (cyclohexane): 2063, 2025, 2014, 1984, 1955, 1941, 1917 (all CO) [9]

23 $(\mu-H)Os_3(CO)_8P(CH_3)_2C_6H_5(\mu_3-\eta^2-NHC_5H_4N)$

by thermolysis of $(\mu-H)Os_3(CO)_9P(CH_3)_2C_6H_5$-$(\mu-\eta^2-NHC_5H_4N)$ (see "Organoosmium Compounds" B 4b, in preparation) in refluxing octane for 2 h, followed by chromatographic workup (49%), along with equal amounts of $(\mu-H)Os_3(CO)_9(\mu_3-\eta^2-NHC_5H_4N)$ (Section 3.1.1.7.3.3) [10]

yellow crystals [10]

^1H NMR (CDCl$_3$, 27 °C): 7.4 to 7.0 (C$_6$H$_5$), 8.46, 6.43, 5.05 (H-6, H-5, and H-3 of C$_5$H$_4$), 2.36, 2.21 (d's, both CH$_3$), -14.70 (d, OsH; J(P,H) = 11); H-4 resonance of C$_5$H$_4$ overlapping with C$_6$H$_5$; no resonance for NH given [10]

IR (cyclohexane): 2065, 2027, 1993, 1987, 1965, 1956, 1943, 1935 (all CO) [10]

compounds with μ-PR$_2$ or μ_3-PR ligands

*24 $(\mu-H)_2Os_3(CO)_8\{\mu-P(C_4H_9-t)_2\}_2$

from Os$_3$(CO)$_{12}$ and PH(C$_4$H$_9$-t)$_2$ (1:3) in refluxing toluene for 12 h, followed by extraction with hexane and column chromatography with hexane and toluene as eluant (50%), along with 24% of Os$_3$(CO)$_{10}\{PH(C_4H_9-t)_2\}_2$ ("Organoosmium Compounds" B 4b, in preparation); a reaction time of 24 h led to a 50% yield, along with 15.4% of $(\mu-H)_2Os_3(CO)_6\{\mu-P(C_4H_9-t)_2\}_2PH(C_4H_9-t)_2$ (Section 3.1.1.8.1) [37]

by carbonylation of $(\mu-H)_2Os_3(CO)_6$-$\{\mu-P(C_4H_9-t)_2\}_2PH(C_4H_9-t)_2$ under 1 atm of CO in hexane [37]

Table 9 (continued)

No. compound	method of preparation (yield) properties and remarks
	yellow-orange, air-stable prisms (from hexane by slow cooling to −10 to −30 °C); m.p. 203 to 205 °C [37] ^{1}H NMR (C_6D_6): 1.31 (m, C_4H_9-t; J(P,H) = 14), −17.50 (t, OsH; J(P,H) = 11.5) [37] ^{31}P {^{1}H} NMR (C_6D_6): 147.00 (s) [37] IR (Nujol mull): 2040, 2000, 1990, 1955, 1941, 1919, 1242, 1150, 1080, 1000, 782, 700, 650, 580 (no assignments given) [37]
*25 $(\mu$-H)$_2$Os$_3$(CO)$_8${μ-P(C_6H_5)$_2$}$_2$	from $(\mu$-H)Os$_3$(CO)$_9${μ-P(C_6H_5)$_2$}PH(C_6H_5)$_2$ (see "Organoosmium Compounds" B 4b, in preparation) in CH_2Cl_2 for 3 h, followed by TLC with hexane/CH_2Cl_2 as eluant (67%) [36] by thermolysis of Os$_3$(CO)$_{10}${PH(C_6H_5)$_2$}$_2$ ("Organoosmium Compounds" B 4b, in preparation) in decahydronaphthalene at ca. 160 °C for 4 h, followed by column chromatography on Florisil with benzene/heptane (1:1) as eluant (60%), along with 6% of $(\mu$-H)Os$_3$(CO)$_{10}${μ-P(C_6H_5)$_2$} ("Organoosmium Compounds" B 4b, in preparation) [20] pale yellow crystals (from toluene) [20] ^{1}H NMR (CDCl$_3$ or CD$_2$Cl$_2$): 8.0 to 7.3 (m, C_6H_5), −17.31 (t, OsH; J(P,H) = 13.7) [36]; (CDCl$_3$): 7.3 to 7.95 (m, C_6H_5), −17.30 (t, OsH; ^{2}J(P,H) = 13.70) [20] ^{31}P {^{1}H} NMR (CDCl$_3$): 63.80 [20] IR (hexane): 2082, 2048, 2022, 2013, 1983, 1961 (all CO) [36]; (cyclohexane): 2078, 2043, 2017, 2008, 1979, 1958 (all CO) [20] EI mass spectrum: [M]$^+$ [36] attempted carbonylation at 80 °C for 96 h failed [20]
26 $(\mu$-H)$_2$Os$_3$(CO)$_8$(μ_3-PC$_6$H$_5$)PH$_2$C$_6$H$_5$ C$_6$H$_5$ | P (CO)$_3$Os ———|——— Os(CO)$_3$ H— Os —H C$_6$H$_5$H$_2$P CO CO	from $(\mu$-H)$_2$Os$_3$(CO)$_9$(μ_3-PC$_6$H$_5$) and PH$_2$C$_6$H$_5$ in refluxing n-butyl ether for 1 h, followed by TLC (9%) [8] yellow, air-stable product [8] IR (cyclohexane): 2075, 2043, 2039, 1999, 1987, 1981, 1967 (all CO); (CHCl$_3$): 1440 (PC$_6$H$_5$), 2340 (PH) [8] EI mass spectrum: [M]$^+$, [M−n CO]$^+$; n=1 to 8 [8]

References on pp. 155/6

Table 9 (continued)

No. compound	method of preparation (yield) properties and remarks

compound with a P- and N-bonded bidentate ligand

*27 $(\mu\text{-H})Os_3(CO)_8\{\mu_3\text{-}\eta^2\text{-}P(C_6H_5)_2NC_6H_5\}P(C_6H_5)_2NHC_6H_5$

from $Os_3(CO)_{12}$ and cyclo-
$\{N(C_6H_5)P(C_6H_5)_2N(C_6H_5)BH_2\}_2$ (1:1) in
CH_3CN at 130 °C for 15 h in a pressure
vessel; workup by repeated TLC with
CH_2Cl_2/cyclohexane (40:60) and
CH_2Cl_2/cyclohexane (5:95) as eluants (9%),
along with 14% of $Os_3(CO)_{11}P(C_6H_5)_2$-
NHC_6H_5 and 11% of $Os_3(CO)_{10}\{P(C_6H_5)_2$-
$NHC_6H_5\}_2$ (both "Organoosmium Compounds" B 4b, in preparation) [32]
by thermolysis of $Os_3(CO)_{10}\{P(C_6H_5)_2NHC_6H_5\}_2$
in CH_3CN at 130 °C for 15 h in a pressure
vessel; workup as before (32%) [32]
pale yellow, air-stable crystals (from
CH_2Cl_2/pentane) [32]
^1H NMR ($CDCl_3$): 7.70 to 5.84 (m, C_6H_5), 3.18
(d, NH; J(P,H) = 19), -13.50 (t, OsH;
J(P,H) = 12) [32]
^{31}P NMR ($CDCl_3$): 25.4, 14.2 (d's; J(P,P) = 22);
no assignments given [32]
IR (CH_2Cl_2): 2074, 2030, 1993, 1967, 1957, 1935
(all CO); (KBr): 3380 (NH) [32]
mass spectrum: $[M]^+$, $[M-n\ CO]^+$, n = 1 to 8
[32]

compounds of the type $Os_3(CO)_8(\mu\text{-}\eta^2\text{-}D\text{-}D)_2$ and an ion derived therefrom

*28 $Os_3(CO)_8(\mu\text{-}\eta^2\text{-dppm})_2$

IV (14%), along with 52% of
$Os_3(CO)_9(\mu\text{-}\eta^2\text{-dppm})$ and 1% of
$Os_3(CO)_{10}(\mu\text{-}\eta^2\text{dppm})$ (both "Organoosmium
Compounds" B 4b, in preparation) [23]
from $Os_3(CO)_{10}(\mu\text{-}\eta^2\text{-dppm})$ and dppm (1:1) in
refluxing toluene for 2 h; workup by TLC
with CH_2Cl_2/hexane (1:2) as eluant (68%)
[43]
from $(\mu\text{-H})_2Os_3(CO)_9(\mu_3\text{-S})$ ("Organoosmium
Compounds" B 3, 1994, p. 191) and dppm
(1:3) in refluxing octane for 4 h, followed by
TLC with light petroleum/CH_2Cl_2 (2:1) as
eluant (11%), along with 15% of
$(\mu\text{-H})_2Os_3(CO)_7(\mu\text{-}\eta^2\text{-dppm})(\mu_3\text{-S})$ (Section
3.1.1.8.2) [40]
orange crystals (from hexane/CH_2Cl_2 at
-20 °C) [43]

Table 9 (continued)

No. compound	method of preparation (yield) properties and remarks

<table>
<tr><td></td><td>^{1}H NMR (CD$_2$Cl$_2$): 4.84 (t, CH$_2$; J(P,H)=9.8) [23]; (CDCl$_3$): 7.35 (m, C$_6$H$_5$), 4.84 (t, CH$_2$; J(P,H)=9.8) [43]
^{31}P {^{1}H} NMR (CDCl$_3$, vs. internal P(OCH$_3$)$_3$): −161.2, −164.7 (both m's) [43]; (CD$_2$Cl$_2$ vs. external H$_3$PO$_4$): AA'BB' spectrum centred at −22.2 [23]
IR (CH$_2$Cl$_2$): 2047, 1989, 1962, 1937, 1895, 1887 (all CO) [23]; similar values in CHCl$_3$ [43]
FAB mass spectrum (2,4-di-t-butylphenol-glycerol): [M]$^+$ [23]</td></tr>
<tr><td>29 [(μ-H)Os$_3$(CO)$_8$(μ-η^2-dppm)$_2$]$^+$</td><td>by protonation of No. 28 with a 10-fold excess of CF$_3$CO$_2$H in CDCl$_3$ in an NMR tube (quantitative; not isolated) [43]
^{1}H NMR (CDCl$_3$, both at ca. +20 °C and −50 °C): 7.17 (m, C$_6$H$_5$), 4.94 (t, CH$_2$; J(P,H)=9.0), −19.42 (t, OsH; J(P,H)=14.4); the hydride is coupled to two equivalent ^{31}P nuclei indicating that μ-H bridges the not by dppm bridged Os−Os edge; the coupling also suggests a transoid relationship with the two ^{31}P nuclei [43]
^{31}P {^{1}H} NMR (CDCl$_3$, vs. internal P(OCH$_3$)$_3$, ca. +20 °C and −50 °C): −166.0, −173.9 (both m's); the shown AA'BB' pattern is consistent with the proposed structure [43]</td></tr>
<tr><td>30 Os$_3$(CO)$_8$(μ-η^2-dppen)$_2$</td><td>IV (60%) [24]
orange in solution [24]
^{31}P {^{1}H} NMR (CD$_2$Cl$_2$): −0.98, −3.84 (d's, each probably 2 P; J(P,P)=96) [24]
IR (CH$_2$Cl$_2$): 2040, 1985, 1960, 1935, 1897 (all CO) [24]</td></tr>
</table>

31 Os$_3$(CO)$_8$\{μ-η^2-(C$_6$H$_5$)$_2$P(3,3,4,4-C$_4$F$_4$-cyclo)P(C$_6$H$_5$)$_2$\}$_2$

structure only proposed
IV (22%), along with 15% of Os$_2$(CO)$_6$(C$_6$H$_5$)$_2$P(3,3,4,4-C$_4$F$_4$-cyclo)-P(C$_6$H$_5$)$_2$; slightly better yields from a 1:3 starting mole ratio [3]
dark red crystals (from CH$_2$Cl$_2$/hexane at 0 °C); only soluble in fairly polar solvents [3]
IR (CH$_2$Cl$_2$): 2064, 2008, 2002, 1982, 1950, 1932 (all CO) [3]
EI mass spectrum: [M]$^+$ not observed; only peaks corresponding to [Os$_3$(CO)$_n$]$^{m+}$ (n=0

References on pp. 155/6

Table 9 (continued)

No. compound	method of preparation (yield) properties and remarks

31 (continued)   to 12, m = 1 and 2), probably formed by
                                       rearrangement in the mass spectrometer [3]

32 $Os_3(CO)_8\{\mu-\eta^2-(CH_3O)_2PN(C_2H_5)P(OCH_3)_2\}_2$

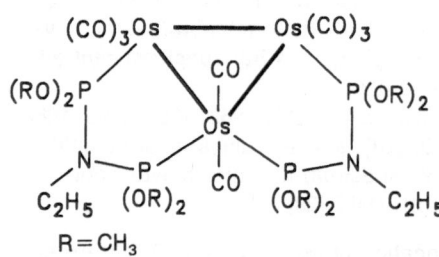

R = CH₃

no preparation reported in [38]
reformed from No. 41 by treatment with an excess of I⁻ [38]
reaction with Cl₂, Br₂, or I₂ resulted in Nos. 39,
40, and 41, respectively; see Preparation
Method III [38]

33 $Os_3(CO)_8\{\mu-\eta^2-(i-C_3H_7O)_2PN(C_2H_5)P(OC_3H_7-i)_2\}_2$
    as No. 32, but R = i-C₃H₇

no preparation reported
reaction with Cl₂, Br₂, or I₂ resulted in Nos. 42,
43, and 44, respectively; see Preparation
Method III [38]

compounds of the type $(\mu-H)_2Os_3(CO)_8(\mu-\eta^2-D-D)$

*34 $(\mu-H)_2Os_3(CO)_8(\mu-\eta^2\text{-dppm})$

V (67 to 84%) [22, 34]
by hydrogenation of $(\mu-H)Os_3\{\mu_3-\eta^3-$
$C_6H_4P(C_6H_5)CH_2P(C_6H_5)_2\}(CO)_8$ ("Organoosmium Compounds" B 6, 1993, p. 116) in
toluene at 80 °C for 0.5 h, followed by crystallization upon cooling of the reaction mixture (75%) [18]
large red, plate-like crystals [18, 22, 34] (from
CH₂Cl₂/ether [34] or CH₂Cl₂/CH₃OH [22]);
m.p. 235 to 238 °C [22]
¹H NMR (CDCl₃): 7.30 (m, C₆H₅), 4.15 (t, CH₂;
J(P,H) = 10.5), −10.31 (t, OsH; J(P,H) = 10.5)
[22, 34]; (CDCl₃): 7.27 (m, C₆H₅), 4.23 (t,
CH₂; J(P,H) = 10.5), −10.22 (t, OsH; J(P,H) =
10.25) [18]
³¹P {¹H} NMR (CDCl₃): −0.53 [18]
IR (cyclohexane): 2076, 2013, 1995, 1974, 1959
(all CO) [22]; (CH₂Cl₂): 2066, 2004, 1982,
1954, 1943 (all CO) [34]; (CH₂Cl₂): 2068,
2005, 1982, 1954, 1943 (all CO) [18]
EI mass spectrum: [M]⁺ [22]

35 $(\mu-H)_2Os_3(CO)_8(\mu-\eta^2\text{-dppe})$

V (28%) [34]
by hydrogenation of $(\mu-H)Os_3\{\mu-\eta^3-$
$C_6H_4P(C_6H_5)(CH_2)_2P(C_6H_5)_2\}(CO)_9$ ("Organoosmium Compounds" B 5, 1994, p. 267) in

Table 9 (continued)

No. compound	method of preparation (yield) properties and remarks
	refluxing cyclohexane for 7 h under 1 atm of H_2; workup by TLC with CH_2Cl_2/light petroleum (1:10) as eluant (31%), along with 10% of $(\mu-H)_3Os_3\{\mu-\eta^3-C_6H_4P(C_6H_5)-(CH_2)_2P(C_6H_5)_2\}(CO)_8$ and 11% of $(\mu-H)_2Os_3-\{\mu_3-\eta^3-C_6H_4P(C_6H_5)(CH_2)_2P(C_6H_5)_2\}(CO)_8$ (both "Organoosmium Compounds" B 5, 1994, p. 267 and p. 321) [35] purple, slightly impure crystals (from CH_2Cl_2/hexane) [34] ^1H NMR (CDCl$_3$): 7.47 (m, C_6H_5), 2.76 (m, CH_2), -10.74 (t, OsH; J(P,H)=9.6) [34] IR (CH$_2$Cl$_2$): 2063, 2007, 1981, 1949 (all CO) [34]
36 $(\mu-H)_2Os_3(CO)_8(\mu-\eta^2-dppp)$	V (63%) [34] purple crystals (from ether) [34] ^1H NMR (CDCl$_3$): 7.39 (m, C_6H_5), 2.70, 1.90 (m's, CH_2), -10.15 (t, OsH; J(P,H)=8.2) [34] IR (CH$_2$Cl$_2$): 2066, 2007, 1980, 1948, 1943 (all CO) [34]

compounds of the types $(\mu-H)Os_3(CO)_8(\mu-\eta^2-D-D)(\mu-X)$ and $[Os_3(CO)_8(\mu-\eta^2-D-D)_2(\mu-X)]X$
$(X = halogen)$

37 $(\mu-H)Os_3(CO)_8(\mu-\eta^2-dmpm)(\mu-Cl)$	from $Os_3(CO)_9(\mu-\eta^2-dmpm)NCCH_3$ (see "Organoosmium Compounds" B 4b, in preparation) by treatment with gaseous HCl in refluxing cyclohexane for 1.5 h; workup by TLC [31] EI mass spectrum: $[M]^+$ [31]; no further spectroscopic data given
38 $(\mu-H)Os_3(CO)_8(\mu-\eta^2-dppm)(\mu-Cl)$	from $Os_3(CO)_9(\mu-\eta^2-dppm)NCCH_3$ as described for No. 37 (48%) [31] from $(\mu-H)Os_3(CO)_{10}(\mu-Cl)$ ("Organoosmium Compounds" B 3, 1994, p. 82) and $(C_6H_5)_2PCH_2P(C_6H_5)_2$ [31]; no details given ^1H NMR (CDCl$_3$): 7.8 to 7.1 (m, C_6H_5), 5.97 (ddd, 1 H of CH_2; J(H,H)=15.8, J(H,P)=10.9, 8.6), 4.46 (dt, 1 H of CH_2; J(H,H)=15.8, J(H,P)=12.0), -12.85 (d, OsH; J(H,P)=32.2) [31] IR (cyclohexane): 2079, 2033, 2007, 1997, 1988, 1968, 1956, 1933 (all CO) [31]

References on pp. 155/6

Table 9 (continued)

No. compound	method of preparation (yield) properties and remarks

39 $[Os_3(CO)_8\{\mu-\eta^2-(CH_3O)_2PN(C_2H_5)P(OCH_3)_2\}_2(\mu-Cl)]Cl$

III [38]

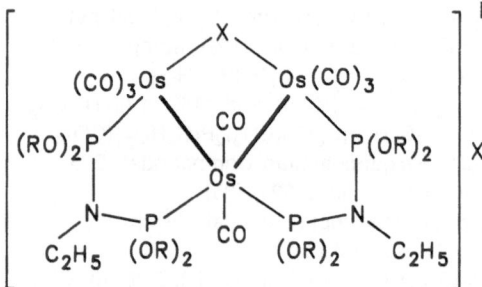

R=CH₃, X=Cl

40 $[Os_3(CO)_8\{\mu-\eta^2-(CH_3O)_2PN(C_2H_5)P(OCH_3)_2\}_2(\mu-Br)]Br$
as No. 39, R=CH₃, X=Br   III [38]

41 $[Os_3(CO)_8\{\mu-\eta^2-(CH_3O)_2PN(C_2H_5)P(OCH_3)_2\}_2(\mu-I)]I$
as No. 39, R=CH₃, X=I   III [38]
treatment with an excess of I⁻ reformed
No. 32; expected **Os₃(CO)₈{μ-η²-
(CH₃O)₂PN(C₂H₅)P(OCH₃)₂}₂I₂** having a linear
Os₃ core and two terminally coordinated
iodide ligands was not observed [38]

42 $[Os_3(CO)_8\{\mu-\eta^2-(i-C_3H_7O)_2PN(C_2H_5)P(OC_3H_7-i)_2\}_2(\mu-Cl)]Cl$
as No. 39, R=C₃H₇–i, X=Cl   III [38]

43 $[Os_3(CO)_8\{\mu-\eta^2-(i-C_3H_7O)_2PN(C_2H_5)P(OC_3H_7-i)_2\}_2(\mu-Br)]Br$
as No. 39, R=C₃H₇–i, X=Br   III [38]

44 $[Os_3(CO)_8\{\mu-\eta^2-(i-C_3H_7O)_2PN(C_2H_5)P(OC_3H_7-i)_2\}_2(\mu-I)]I$
as No. 39, R=C₃H₇–i, X=I   III [38]

**compounds of the types (μ-H)Os₃(CO)₈(μ-η²-D-D)(μ-OH) and
(μ-H)Os₃(CO)₈(μ-η²-D-D)(μ-SR) and ions derived therefrom**

*45 (μ-H)Os₃(CO)₈(μ-η²-dppm)(μ-OH)   from (μ-H)Os₃(CO)₁₀(μ-OH) ("Organoosmium
Compounds" B 3, 1994, p. 94) and
(C₆H₅)₂PCH₂P(C₆H₅)₂ in refluxing hexane for
5 h; workup by TLC with CH₂Cl₂/hexane
(1:1) as eluant (70%) [31]
yellow, hexagonal, air-stable plates (from
CH₂Cl₂/C₂H₅OH) [31]
¹H NMR (CDCl₃): 7.4 to 7.3 (m, C₆H₅), 3.78
(dtd, 1 H of CH₂; J(H,H)=13.4, 1.7, J(P,H)=
11.4), 2.69 (dt, 1 H of CH₂; J(H,H)=13.4,

References on pp. 155/6

Table 9 (continued)

No. compound	method of preparation (yield) properties and remarks

$J(P,H) = 10.7$), 0.44 (td, OH; $J(P,H) = 4.3$,
$J(H,H) = 1.5$), -11.68 (t, OsH; $J(P,H) = 11.9$)
[31]
^{31}P {^1H} NMR (CH$_2$Cl$_2$; vs. internal P(OCH$_3$)$_3$):
-152.8 (s) [31]
IR (cyclohexane): 3625 (OH), 2069, 2006, 1999,
1964, 1940 (all CO) [31]

46 (μ-H)Os$_3$(CO)$_8$(μ-η^2-dmpm)(μ-SCH$_3$)

VI [31]
^1H NMR (CDCl$_3$): 4.36 (dt, 1 H of CH$_2$; $J(H,H) = $
14.5, $J(H,P) = 9.9$), 3.38 (dt, 1 H of CH$_2$;
$J(H,H) = 14.5$, $J(H,P) = 11.5$), 2.33 (s, SCH$_3$),
2.12 (d, PCH$_3$; $J(H,P) = 9.6$), 1.92 (d, PCH$_3$;
$J(H,P) = 7.9$), 1.90 (d, PCH$_3$; $J(H,P) = 8.5$), 1.81
(d, PCH$_3$; $J(H,P) = 8.9$), -16.14 (d, OsH;
$J(H,P) = 26.2$) [31]
IR (cyclohexane): 2070, 2020, 1994, 1987, 1972,
1958, 1947, 1927 (all CO) [31]

47 (μ-H)Os$_3$(CO)$_8$(μ-η^2-dmpm)(μ-SC$_2$H$_5$)

VI [31]
^1H NMR (CDCl$_3$): 4.40 (dt, 1 H of PCH$_2$;
$J(H,H) = 14.5$, $J(H,P) = 9.8$), 3.37 (dt, 1 H of
PCH$_2$; $J(H,H) = 14.5$, $J(H,P) = 11.4$), 2.50 (dq,
1 H of SCH$_2$), 2.2 to 2.1 (m, 1 H of SCH$_2$ and
PCH$_3$), 2.0 to 1.7 (m, PCH$_3$), 1.20 (t, CH$_3$ of
SC$_2$H$_5$; $J(H,H) = 7.3$), -16.26 (d, OsH;
$J(H,P) = 26.6$) [31]
^{31}P {^1H} NMR (CH$_2$Cl$_2$; vs. internal P(OCH$_3$)$_3$):
-202.1, -202.0 (d's; $J(P,P) = 35$); no assign-
ments given [31]
IR (cyclohexane): 2070, 2020, 1993, 1987, 1971,
1958, 1947, 1925 (all CO) [31]
EI mass spectrum: [M]$^+$ [31]

48 (μ-H)Os$_3$(CO)$_8$(μ-η^2-dmpm)(μ-SC$_6$H$_5$)

VI [31]
^1H NMR (CDCl$_3$): 7.2 to 7.1 (m, C$_6$H$_5$), 4.43 (dt,
1 H of CH$_2$; $J(H,H) = 14.5$, $J(H,P) = 9.8$), 3.42
(dt, 1 H of CH$_2$; $J(H,H) = 14.5$, $J(H,P) = 11.2$),
2.15 (d, PCH$_3$; $J(H,P) = 9.5$), 2.0 to 1.8 (m,
PCH$_3$), -15.87 (d, OsH; $J(H,P) = 26.7$) [31]
IR (cyclohexane): 2071, 2021, 1995, 1989, 1975,
1959, 1948, 1929 (all CO) [31]

References on pp. 155/6

Table 9 (continued)

No. compound	method of preparation (yield) properties and remarks

49 $(\mu-H)Os_3(CO)_8(\mu-\eta^2-dppm)(\mu-SCH_3)$

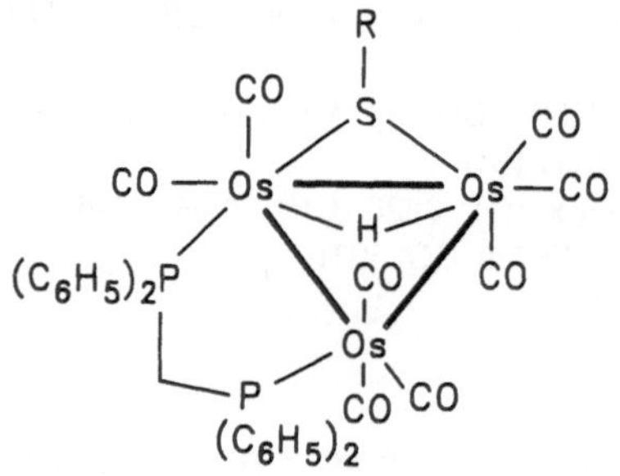

R = CH₃

VI [31]

^1H NMR (CDCl₃): 7.6 to 7.3 (m, C₆H₅), 5.52 (dt, 1 H of CH₂; J(H,H) = 15.7, J(H,P) = 10.1), 4.73 (dt, 1 H of CH₂; J(H,H) = 15.7, J(H,P) = 11.0), 2.11 (s, CH₃), −16.03 (d, OsH; J(H,P) = 29.1) [31]

IR (cyclohexane): 2071, 2027, 1996, 1989, 1973, 1964, 1960, 1953, 1944, 1936, 1929 (all CO) [31]

50 $(\mu-H)Os_3(CO)_8(\mu-\eta^2-dppm)(\mu-SC_2H_5)$
as No. 49, R = C₂H₅

VI (70%) [31]

^1H NMR (CDCl₃): 7.7 to 7.1 (m, C₆H₅), 5.41 (ddd, 1 H of PCH₂; J(H,H) = 15.3, J(H,P) = 11.2, 9.0), 4.85 (dt, 1 H of PCH₂; J(H,H) = 15.3, J(H,P) = 10.3), 2.11 (m, SCH₂), 0.94 (t, CH₃; J(H,H) = 7.3), −16.19 (d, OsH; J(H,P) = 30.8) [31]

^{31}P {^1H} NMR (CH₂Cl₂ vs. internal P(OCH₃)₃): −165.9 (d; J(P,P) = 39), −168.7 (d; J(P,P) = 39); no assignments given [31]

IR (cyclohexane): 2070, 2026, 1996, 1988, 1972, 1963, 1959, 1953, 1944, 1936, 1928 (all CO) [31]

51 $(\mu-H)Os_3(CO)_8(\mu-\eta^2-dppm)(\mu-SC_3H_7-n)$
as No. 49, R = C₃H₇-n

VI (9%) [39]

orange crystals [39]

^1H NMR (CDCl₃): 7.35 (m C₆H₅), 5.12 (m, CH₂ of dppm), 2.05 (t, SCH₂), 1.25 (m, CH₂), 0.85 (t, CH₃), −16.19 (d, OsH; J(P,H) = 29.5, indicating that μ-H is trans to P); spectrum of the hydride region depicted [39]

IR (CH₂Cl₂): 2067, 2027, 1986, 1953, 1927 [39]

protonation with an excess of CF₃CO₂H in CDCl₃ followed by treatment with NH₄PF₆ in CH₃OH led to No. 52 [39]

Table 9 (continued)

No.	compound	method of preparation (yield) properties and remarks

52 [(μ-H)₂Os₃(CO)₈(μ-η²-dppm)(μ-SC₃H₇-n)]PF₆

R = C₃H₇-n

by protonation of No. 51 with a tenfold excess of CF_3CO_2H in $CDCl_3$, followed by treatment with NH_4PF_6 in CH_3OH and precipitation with small amounts of H_2O (59%) [39]
yellow crystals (from CH_2Cl_2/ether) [39]
^1H NMR ($CDCl_3$): 7.35 (m, C_6H_5), 5.14 (m, CH_2 of dppm), 1.83 (m, SCH_2), 1.24 (m, CH_2), 0.83 (t, CH_3), −15.64 (dd, OsH; J(P,H) = 19.04, J(H,H) = 1.3), −18.87 (dd, OsH; J(P,H) = 10.99, J(H,H) = 1.3); spectrum of the hydride region depicted; the magnitudes of the P-H coupling constants indicate that one of the hydrides is trans to the P atom with which it is coupled whereas the other one is cis to the P atom [39]
IR (CH_2Cl_2): 2126, 2097, 2066, 2059, 2023, 1995 (all CO) [39]

53 (μ-H)Os₃(CO)₈(μ-η²-dppm)(μ-SC₄H₉-t)
as No. 49, R = C₄H₉-t

VI [31]
IR (cyclohexane): 2069, 2025, 1996, 1989, 1966, 1960, 1954, 1946, 1933, 1929 (all CO) [31]

54 (μ-H)Os₃(CO)₈(μ-η²-dppm)(μ-SC₆H₅)
as No. 49, R = C₆H₅

VI [31]
from $Os_3(CO)_{10}(μ-η^2$-dppm) ("Organoosmium Compounds" B 4b, in preparation) and C_6H_5SH (1:3) in refluxing toluene for 3 h, followed by TLC with light petroleum/CH_2Cl_2 (10:3) as eluant (49%) [40]
orange crystals (from hexane/CH_2Cl_2) [40]
^1H NMR ($CDCl_3$, 27 °C): 7.51 (m, C_6H_5), 5.75, 4.61 (m's, each 1 H of CH_2), −15.75 (d, OsH; J = 31.0) [40]
^{31}P {^1H} NMR ($CDCl_3$): −29.02 (s), −28.42 (s) [40]
IR (CH_2Cl_2): 2065, 2024, 1989, 1934, 1921 (all CO) [40]; (CH_2Cl_2): 2069, 2024, 1994, 1972, 1956, 1946, 1927 (all CO) [31]
protonation with an excess of CF_3CO_2H followed by treatment with NH_4PF_6 in CH_3OH resulted in No. 55 [40]

55 [(μ-H)₂Os₃(CO)₈(μ-η²-dppm)(μ-SC₆H₅)]PF₆
as No. 52, R = C₆H₅

by protonation of No. 54 with a tenfold excess of CF_3CO_2H in $CDCl_3$, followed by treatment

Table 9 (continued)

No. compound	method of preparation (yield) properties and remarks
55 (continued)	with NH$_4$PF$_6$ in CH$_3$OH and precipitation with small amounts of H$_2$O (52%) [40] pale yellow crystals (from CH$_2$Cl$_2$/ether) [40] ^{1}H NMR (CDCl$_3$): 7.42 (m, C$_6$H$_5$), 5.30 (m, CH$_2$), −15.48 (d, OsH; J=22), −18.52 (d, OsH; J=11) [40] IR (CH$_2$Cl$_2$): 2130, 2097, 2063, 2059, 2015, 2004, 1995 (all CO) [40]
56 (μ-H)Os$_3$(CO)$_8$(μ-η^2-dppm)(μ-SC$_6$H$_4$CH$_3$-4) as No. 49, R=C$_6$H$_4$CH$_3$-4	from Os$_3$(CO)$_{10}$(μ-η^2-dppm) ("Organoosmium Compounds" B 4b, in preparation) and 4-CH$_3$C$_6$H$_4$SH as described for No. 54 (57%) [40] orange crystals (from hexane/CH$_2$Cl$_2$) [40] ^{1}H NMR (CDCl$_3$, 27 °C): 7.42 (m, C$_6$H$_5$ and C$_6$H$_4$), 4.81, 3.53 (m's, each 1 H of CH$_2$), 2.35 (s, CH$_3$), −15.76 (d, OsH; J=29.5) [40] ^{31}P $\{^1$H$\}$ NMR (CDCl$_3$): −28.85 (s), −28.66 (s) [40] IR (CH$_2$Cl$_2$): 2065, 2023, 1993, 1969, 1954, 1920 (all CO) [40] protonation with CF$_3$CO$_2$H followed by treatment with NH$_4$PF$_6$ in CH$_3$OH resulted in No. 57 [40]
57 [((μ-H)$_2$Os$_3$(CO)$_8$(μ-η^2-dppm)(μ-SC$_6$H$_4$CH$_3$-4)]PF$_6$ as No. 52, R=C$_6$H$_4$CH$_3$-4	by protonation of No. 56 with CF$_3$CO$_2$H as described for No. 55 (60%) [40] yellow crystals (from CH$_2$Cl$_2$/ether) [40] ^{1}H NMR (CDCl$_3$): 7.31 (m, C$_6$H$_5$ and C$_6$H$_4$), 5.27 (m, CH$_2$), 2.35 (s, CH$_3$), −15.40 (d, OsH; J=21), −18.50 (d, OsH; J=11) [40] IR (CH$_2$Cl$_2$): 2131, 2099, 2065, 2059, 2018, 2006, 1994 (all CO) [40]

compounds of the type Os$_3$(CO)$_8$(μ-η^2-D-D)D$_2$

*58 Os$_3$(CO)$_8$(μ-η^2-dppm){P(OCH$_3$)$_3$}$_2$

D=P(OCH$_3$)$_3$

isomeric with No. 59 [41]
I (22%), along with No. 59 (16%) [41]
^{31}P $\{^1$H$\}$ NMR (no solvent given): 105.7 (s, P(OCH$_3$)$_3$), −29.0 (s, μ-η^2-P) [41]
IR (CH$_2$Cl$_2$): 2043, 1985, 1960, 1920 [41]
mass spectrum: [M]$^+$ [41]

References on pp. 155/6

Table 9 (continued)

No. compound	method of preparation (yield) properties and remarks

59 $Os_3(CO)_8(\mu-\eta^2-dppm)\{P(OCH_3)_3\}_2$

D = P(OCH_3)_3

isomeric with No. 58 [41]
for formation, see No. 58 [41]
^{31}P $\{^1H\}$ NMR (no solvent given): 108.6, 98.9
(d's, each 1 P(OCH_3)_3; J(P,$\mu-\eta^2$-P) = 3.7,
5.5), −22.4 (d, 1 P of $\mu-\eta^2$-P;
$^2J(\mu-\eta^2$-P,P) = 51.4), −27.4 (dt, 1 P of
$\mu-\eta^2$-P; $^2J(\mu-\eta^2$-P,P) = 51.4,
J($\mu-\eta^2$-P,P(OCH_3)_3) = 3.7, 5.5) [41]
IR (CH_2Cl_2): 2052, 1989, 1960, 1918, 1895 [41]
attempted isomerization into No. 58 in toluene
at 50 °C for 12 h failed completely, based on
^{31}P NMR spectra [41]

60 $Os_3(CO)_8(\mu-\eta^2-dppm)\{P(OC_3H_7-i)_3\}_2$
as No. 58, D = P(OC_3H_7-i)_3

I (no yield given; impure) [41]
^{31}P $\{^1H\}$ NMR (no solvent given): 95.6 (s,
P(OC_3H_7-i)_3), −30.5 (s, $\mu-\eta^2$-P) [41]

61 $Os_3(CO)_8(\mu-\eta^2-dppm)\{P(OC_4H_9-n)_3\}_2$
as No. 58, D = P(OC_4H_9-n)_3

I (no yield given; impure) [41]
^{31}P $\{^1H\}$ NMR (no solvent given): 98.4 (s,
P(OC_4H_9-n)_3), −29.4 (s, $\mu-\eta^2$-P) [41]

62 $Os_3(CO)_8(\mu-\eta^2-dppm)\{P(OC_6H_5)_3\}_2$
as No. 58, D = P(OC_6H_5)_3

I (no yield given; impure) [41]
^{31}P $\{^1H\}$ NMR (no solvent given): 75.9 (t,
P(OC_6H_5)_3), −29.0 (t, $\mu-\eta^2$-P);
J(P,$\mu-\eta^2$-P) = 8.8 [41]

63 $Os_3(CO)_8(\mu-\eta^2-dppm)\{P(C_2H_5)_3\}_2$
as No. 58, D = P(C_2H_5)_3

I (33%) [41]
^{31}P $\{^1H\}$ NMR (no solvent given): 52.5 (s,
P(C_2H_5)_3), −32.1 (s, $\mu-\eta^2$-P) [41]

64 $Os_3(CO)_8(\mu-\eta^2-dppm)\{P(C_4H_9-n)_3\}_2$
as No. 58, D = P(C_4H_9-n)_3

I (46%) [41]
^{31}P $\{^1H\}$ NMR (no solvent given): 48.7 (s,
P(C_4H_9-n)_3), −30.6 (s, $\mu-\eta^2$-P) [41]
IR (CH_2Cl_2): 2035, 1970, 1950, 1900, 1875 [41]
mass spectrum: [M]$^+$ [41]

*65 $Os_3(CO)_8(\mu-\eta^2-dppm)\{P(C_6H_5)_3\}_2$
as No. 59, D = P(C_6H_5)_3

I (27%) [41]
for formation, see No. 58 [41]
^{31}P $\{^1H\}$ NMR (no solvent given): 0.4 (d,
1 P(C_6H_5)_3; J(P,$\mu-\eta^2$-P) = 7.8), −9.4 (s,
1 P(C_6H_5)_3), −26.9 (dd, 1 P of $\mu-\eta^2$-P;
$^2J(\mu-\eta^2$-P,P) = 53.1,
J($\mu-\eta^2$-P,P(C_6H_5)_3) = 7.8), −28.7 (d, 1 P of
$\mu-\eta^2$-P; $^2J(\mu-\eta^2$-P,P) = 53.1) [41]
IR (CH_2Cl_2): 2040, 1985, 1960, 1925, 1880 [41]
mass spectrum: [M]$^+$ [41]

References on pp. 155/6

Table 9 (continued)

No. compound	method of preparation (yield) properties and remarks
66 $Os_3(CO)_8(\mu-\eta^2-dppen)\{P(OCH_3)_3\}_2$	probably prepared from $(\mu-H)Os_3\{\mu_3-\eta^3-C_6H_4P(C_6H_5)C(=CH_2)P(C_6H_5)_2\}(CO)_8$ ("Organoosmium Compounds" B 6, 1993, p. 119) by treatment with $P(OCH_3)_3$ under the conditions of Preparation Method I [41, 42] structural determinations indicated that the title compound is essentially identical to No. 58 [41, 42]; no spectroscopic and structural data reported

*Further information:

$Os_3(CO)_8P(CH_3)_2C_6H_5(\mu_3-S)_2$ (Table **9**, No. **17**) crystallizes in the triclinic space group $P\bar{1}-C_i^1$ (No. 2) with a = 9.361(2), b = 11.346(1), c = 12.221(2) Å, α = 71.65(1)°, β = 71.63(2)°, γ = 79.67(1)°; Z = 2, D_c = 2.84 g/cm³. The structure is shown in **Fig. 50**. The cluster consists of an open Os_3 framework with two significantly differing Os-Os bond distances of 2.856 and 2.770 Å, and a non-bonding interaction of 3.713 Å. The $P(CH_3)_2C_6H_5$ ligand occupies

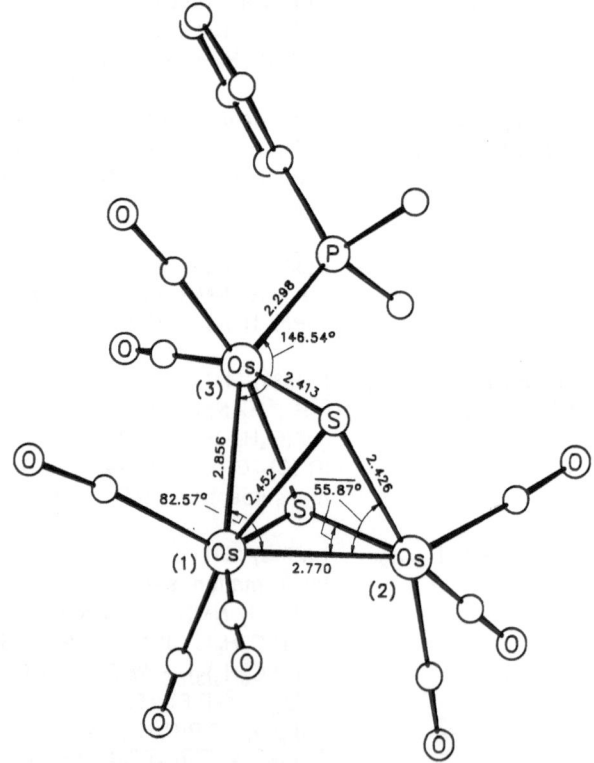

Fig. 50. Molecular structure of $Os_3(CO)_8P(CH_3)_2C_6H_5(\mu_3-S)_2$ (No. 17) with selected bond distances (in Å) and bond angles [11].

References on pp. 155/6

an equatorial coordination site on Os(3) trans to the elongated Os(1)-Os(3) bond. The lengthening of the Os(1)-Os(3) bond is attributed to a trans influence of the phosphane ligand
which is largely σ inductive in character and which leads to a weakening of the Os(1)-Os(3)
bond directly opposite to the phosphane and a strengthening of the Os(1)-Os(2) bond [11].

$Os_3(CO)_8P(OCH_3)_3(NO)_2$ (Table **9**, No. **19**) crystallizes in the monoclinic space group
$P2_1/c-C_{2h}^5$ (No. 14) with a = 15.653(5), b = 8.255(7), c = 17.263(5) Å, β = 105.15(5)°; Z = 4,
D_c = 3.019 g/cm³. The structure is shown in **Fig. 51**. The structure closely relates to that
of $Os_3(CO)_{12}$ in which the four CO groups of one Os center were substituted by an equatorial
phosphite ligand and two terminal NO ligands, which are coordinated approximately equidis-
tant on opposite sides of the Os_3 plane. The P atom lies 0.14 Å from this plane [6].

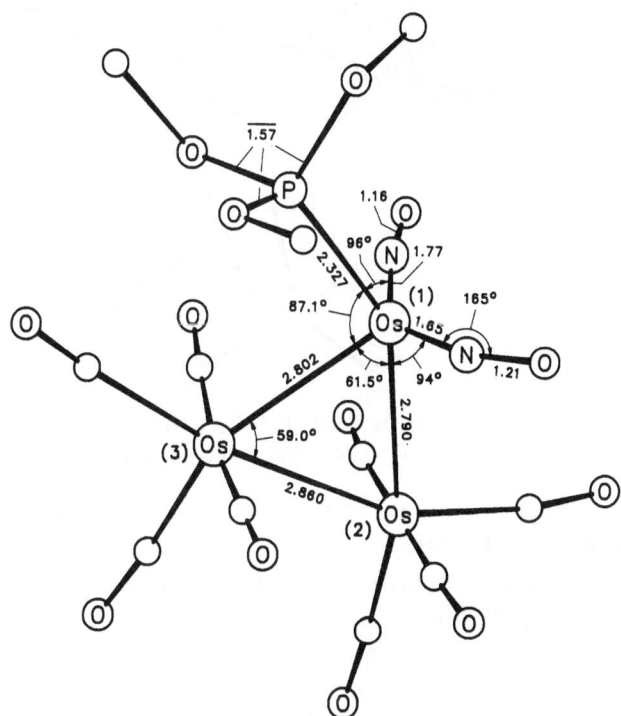

Fig. 51. Molecular structure of $Os_3(CO)_8P(OCH_3)_3(NO)_2$ (No. 19) with selected bond distances
(in Å) and bond angles [6].

Carbonylation in refluxing n-heptane for 30 min led to $Os_3(CO)_9P(OCH_3)_3(\mu-NO)_2$ (see
"Organoosmium Compounds" B 4b, in preparation) [5, 7].

$(\mu-H)_2Os_3(CO)_8\{\mu-P(C_4H_9-t)_2\}_2$ (Table **9**, No. **24**) crystallizes in the triclinic space group
$P\bar{1}-C_i^1$ (No. 2) with a = 10.993(3), b = 17.565(5), c = 8.774(5) Å, α = 93.27(3)°, β = 105.74(2)°,
γ = 106.93(2)°; Z = 2, D_c = 2.340 g/cm³. The structure of the 48-electron cluster is shown
in **Fig. 52**. The coordination geometry of each Os center is roughly octahedral upon ignoring
the metal-metal bonds for Os(2) but including the Os(2)-Os(3) interaction for Os(2) and
Os(3). The two $\mu-P(C_4H_9-t)_2$ groups bridge the Os(1)-Os(2) and Os(1)-Os(3) bond edges
lying above and below the Os_3 plane; the P(1)-Os(1)-P(2) angle amounts to 151.15(9)°.
Although the hydrides were not observed crystallographically, they are believed to bridge
the same Os-Os distances but on opposite to the $\mu-P(C_4H_9-t)_2$ ligands [37].

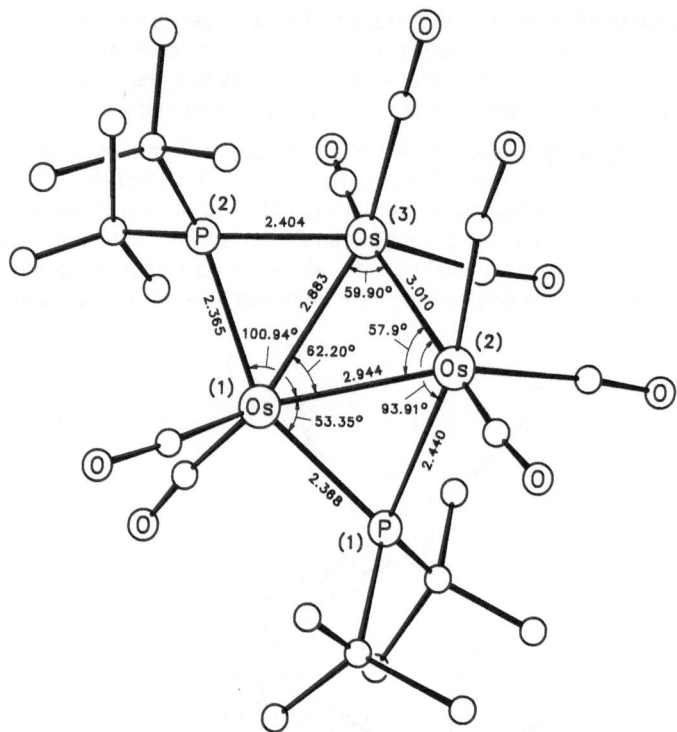

Fig. 52. Molecular structure of $(\mu-H)_2Os_3(CO)_8\{\mu-P(C_4H_9-t)_2\}_2$ (No. 24) with selected bond distances (in Å) and bond angles [37].

Thermolysis in n-butyl ether at 143 °C for 24 h yielded $(\mu-H)_2Os_3(CO)_7(\mu_3-PC_4H_9-t)-\{PH(C_4H_9-t)_2\}_2$ (Section 3.1.1.8.2) [37].

Treatment with an excess of $PH(C_4H_9-t)_2$ in THF gave $(\mu-H)_2Os_3(CO)_6\{\mu-P(C_4H_9-t)_2\}_2-PH(C_4H_9-t)_2$ (Section 3.1.1.8.1) [37].

$(\mu-H)_2Os_3(CO)_8\{\mu-P(C_6H_5)_2\}_2$ (Table 9, No. 25) crystallizes in the triclinic space group $P\bar{1}-C_i^1$ (No. 2) with a = 11.184(1), b = 11.991(1), c = 14.657(1) Å, α = 76.17(1)°, β = 84.72(1)°, γ = 70.01(1)°; Z = 2, D_m = 2.24 g/cm³, D_c = 2.246 g/cm³. A perspective view of the skeleton of the molecule is shown in **Fig. 53**. The compound is isostructural with $(\mu-H)_2M_3(CO)_8\{\mu-P(C_6H_5)_2\}_2$ (M = Fe, Ru). The Os(1)–Os(2) and Os(2)–Os(3) bond edges of the nearly equilateral Os₃ triangle are each bridged by a phosphido and a μ-H ligand above and below the Os₃ plane. P(1) occupies an axial site on Os(3), while P(2) is axial on Os(1); P(1) and P(2) are approximately trans to one another, the P(1)–Os(2)–P(2) angle amounts to 139.85(8)°. The stereochemistry of Os(2) is distorted octahedral. Interestingly, the Os(2)–P bond lengths are significantly shorter than the axial Os(1)–P and Os(3)–P distances, suggesting a stronger association of the phosphido bridges with Os(2) and of the hydrides with Os(1) and Os(3) [20].

$(\mu-H)Os_3(CO)_8\{\mu_3-\eta^2-P(C_6H_5)_2NC_6H_5\}P(C_6H_5)_2NHC_6H_5$ (Table 9, No. 27) crystallizes in the rhombohedral space group $R\bar{3}-C_{3i}^2$ (No. 148) with a = 21.143(9) Å, α = 106.89(3)°; Z = 6, D_c = 1.703 g/cm³. The structure is shown in **Fig. 54**. The Os(1)–Os(2) bond of the equilateral Os₃ triangle is symmetrically bridged by the hydride ligand and the nitrogen of the $\mu_3-\eta^2-P(C_6H_5)_2NC_6H_5$ ligand lying on opposite sides of the Os₃ plane [32].

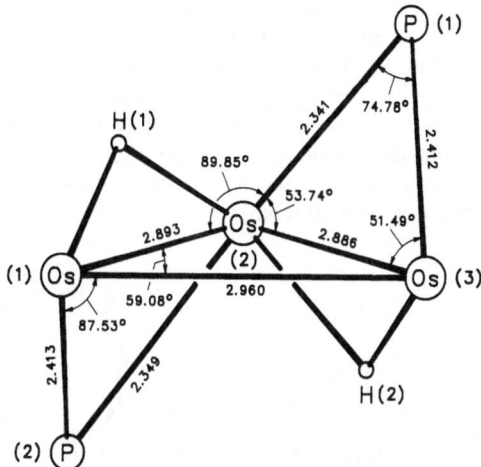

Fig. 53. Perspective view of the skeleton of $(\mu-H)_2Os_3(CO)_8\{\mu-P(C_6H_5)_2\}_2$ (No. 25) with selected bond distances (in Å) and bond angles [20].

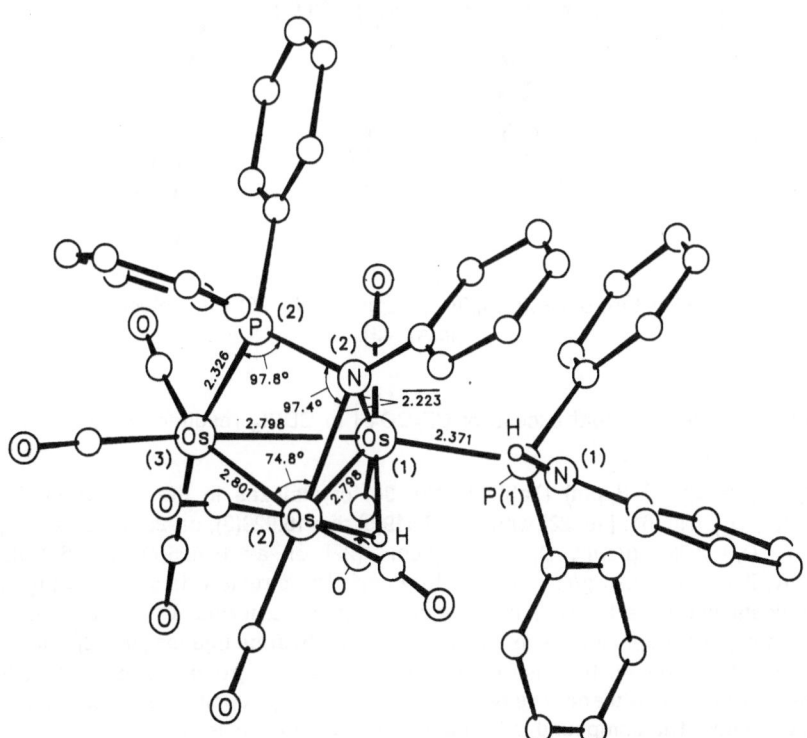

Fig. 54. Molecular structure of $(\mu-H)Os_3(CO)_8\{\mu_3-\eta^2-P(C_6H_5)_2NC_6H_5\}P(C_6H_5)_2NHC_6H_5$ (No. 27) with selected bond distances (in Å) and bond angles [32].

Os$_3$(CO)$_8$(μ-η^2-dppm)$_2$ (Table **9**, No. **28**) crystallizes in the orthorhombic space group Pca2$_1$ – C$_{2v}^5$ (No. 29) with a = 21.398(3), b = 15.684(4), c = 18.219(4) Å; Z = 4, D$_c$ = 1.70 g/cm^3. The structure is shown in **Fig. 55**. The molecule consists of an isosceles Os$_3$ core having two bidentate μ-η^2-dppm ligands occupying equatorial coordination sites on the Os centers. The remaining nonbridged Os–Os bond edge is slightly longer than the bridged Os–Os bonds and is similar to the Os–Os distances in Os$_3$(CO)$_{12}$. The average Os–CO bonds for the axial carbonyl groups are ca. 1.903 Å and for the equatorial carbonyl groups ca. 1.76 Å; average C–O bonds are ca. 1.19 Å [43].

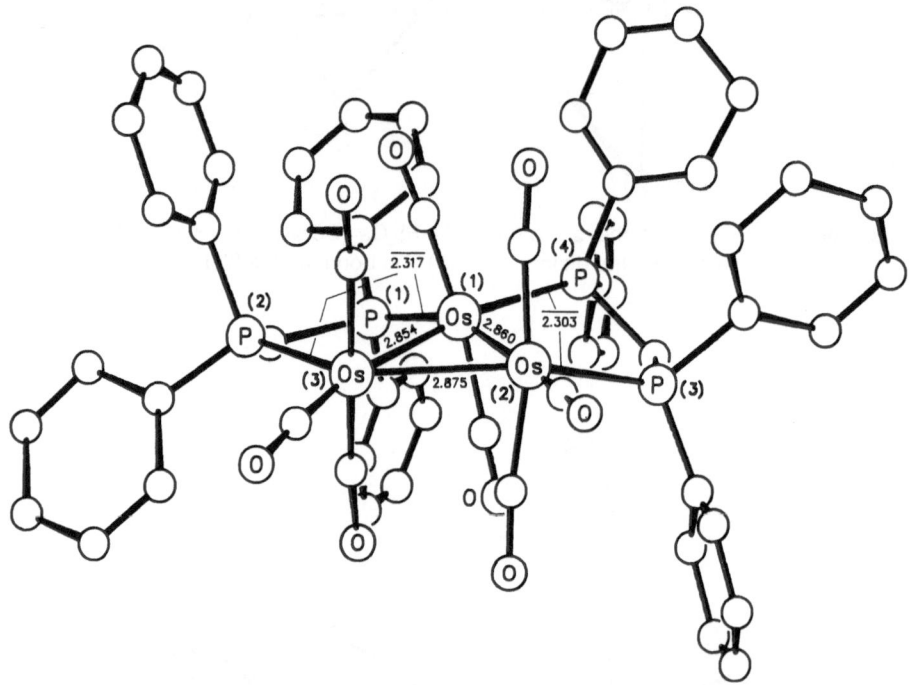

Fig. 55. Molecular structure of Os$_3$(CO)$_8$(μ-η^2-dppm)$_2$ (No. 28) with selected bond distances (in Å) [43].

Protonation with a 10-fold excess of CF$_3$CO$_2$H in CDCl$_3$ resulted in No. 29, based on ^{1}H and ^{31}P NMR spectra [43].

(μ-H)$_2$Os$_3$(CO)$_8$(μ-η^2-dppm) (Table **9**, No. **34**) crystallizes in the orthorhombic space group Pbca – D$_{2h}^{15}$ (No. 61) [18, 22] with a = 16.419(5), b = 16.929(3), c = 25.403(6) Å, α = β = γ = 90°; Z = 8, D$_c$ = 2.222 g/cm^3, D$_m$ = 2.22 g/cm^3 [22] or a = 16.405(7), b = 16.906(2), c = 25.370(3) Å; Z = 8, D$_c$ = 2.24 g/cm^3, D$_m$ = 2.16 g/cm^3. The structure is shown in **Fig. 56** [18]. The significant short Os(1)–Os(2) bond edge indicating a formal Os=Os double bond is bridged by the μ-η^2-dppm moiety as well as by the two hydride ligands [18, 22]; the hydrides were not located directly, but distribution of the carbonyl groups indicates that they lie above and below the Os$_3$ plane. The two P atoms of the μ-η^2-dppm ligand occupy equatorial coordination sites. The compound is formally a 46-electron cluster [18].

(μ-H)Os$_3$(CO)$_8$(μ-η^2-dppm)(μ-OH) (Table **9**, No. **45**) crystallizes in the monoclinic space group P2$_1$/n – C$_{2h}^5$ (No. 14) with a = 12.975(3), b = 11.790(1), c = 23.904(4) Å, β = 105.36(2)°;

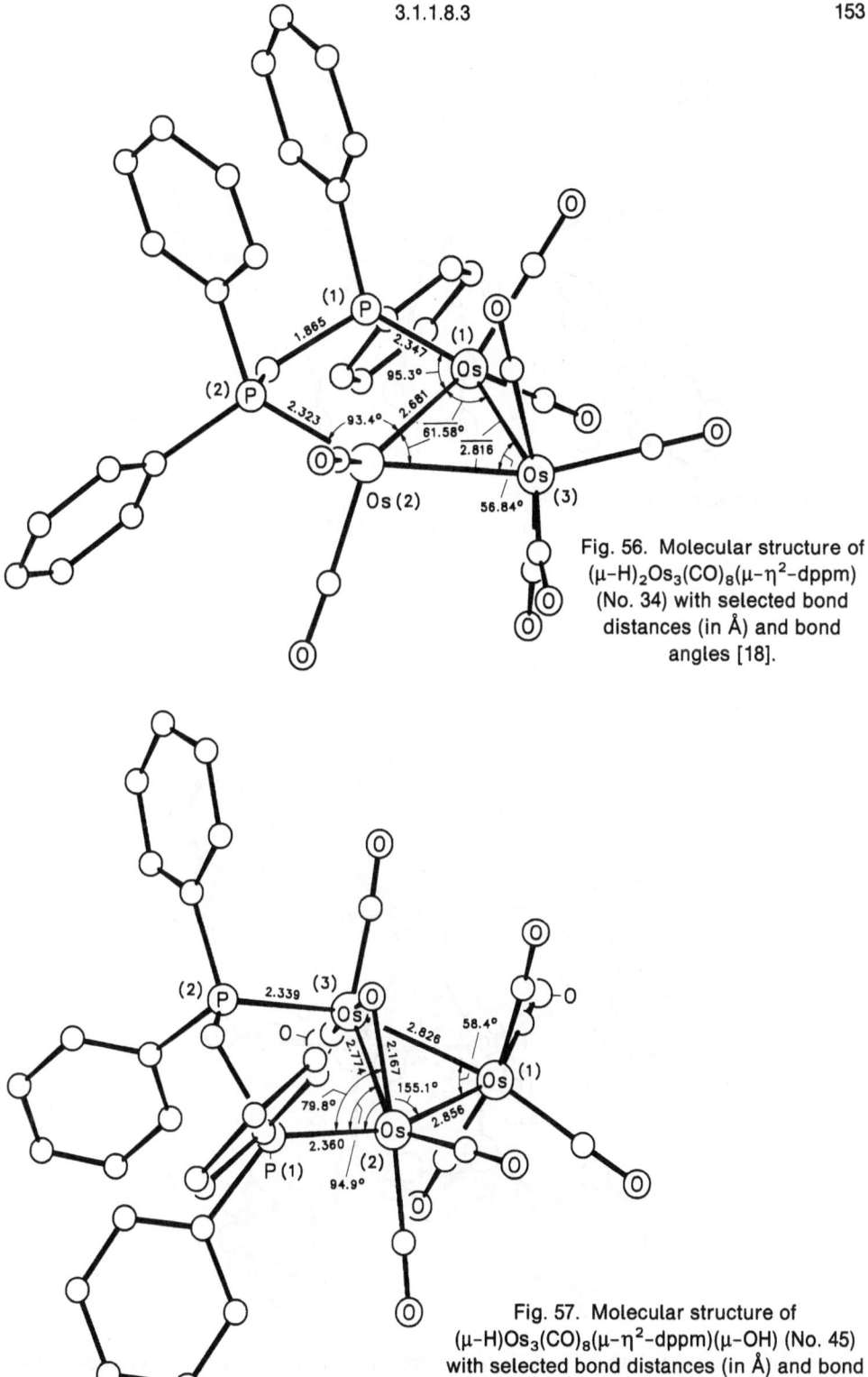

Fig. 56. Molecular structure of $(\mu\text{-}H)_2Os_3(CO)_8(\mu\text{-}\eta^2\text{-dppm})$ (No. 34) with selected bond distances (in Å) and bond angles [18].

Fig. 57. Molecular structure of $(\mu\text{-}H)Os_3(CO)_8(\mu\text{-}\eta^2\text{-dppm})(\mu\text{-}OH)$ (No. 45) with selected bond distances (in Å) and bond angles [31].

References on pp. 155/6

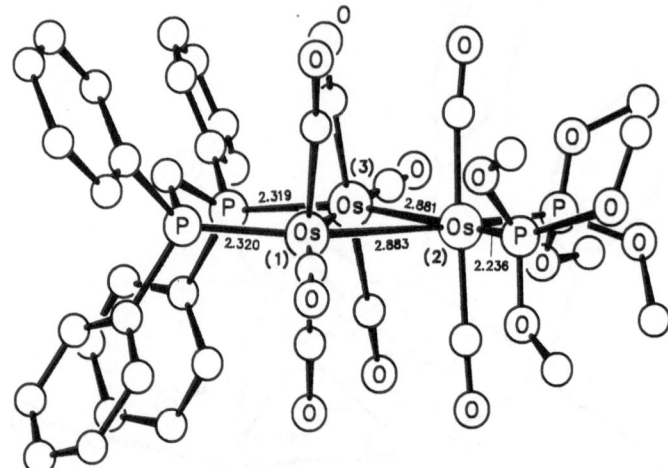

Fig. 58. Molecular structure of $Os_3(CO)_8(\mu-\eta^2-dppm)\{P(OCH_3)_3\}_2$ (No. 58) with selected bond distances (in Å) [41].

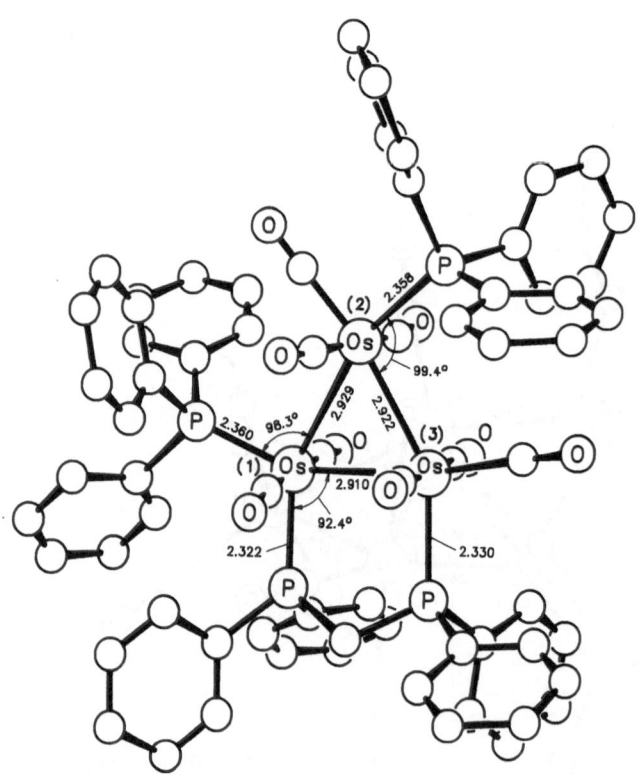

Fig. 59. Molecular structure of $Os_3(CO)_8(\mu-\eta^2-dppm)\{P(C_6H_5)_3\}_2$ (No. 65) with selected bond distances (in Å) and bond angles [41].

$Z = 4$, $D_c = 2.25$ g/cm^3. The structure of the 48-electron cluster molecule is shown in **Fig. 57**. The μ-η^2-dppm ligand, the μ-OH group and supposedly the hydride all bridge the Os(2)-Os(3) edge of the closed Os$_3$ triangle. The two P atoms are essentially in the Os$_3$ plane, μ-OH and μ-H are above and below this plane. The angle between the Os(1)-Os(2)-Os(3) and the Os(2)-Os(3)-P(1)-P(2) plane is only 2.5° [31].

Os$_3$(CO)$_8$(μ-η^2-dppm){P(OCH$_3$)$_3$}$_2$ (Table 9, No. **58**) crystallizes in the monoclinic space group I2/a − C$_{2h}^6$ (No. 15) with a = 22.793(1), b = 10.273(4), c = 38.898(1) Å, β = 91.58(3)°; Z = 8, $D_c = 2.08$ g/cm^3. The structure is shown in **Fig. 58**. The molecule consists of an essentially isosceles Os$_3$ core. The distances of the two unbridged Os-Os bonds amounted to ca. 2.882 Å; the longer bond edge of 2.94 Å is spanned by the μ-η^2-dppm moiety. All P atoms are coordinated in equatorial sites with the two P(OCH$_3$)$_3$ ligands bonded to the same Os center [41].

Os$_3$(CO)$_8$(μ-η^2-dppm){P(C$_6$H$_5$)$_3$}$_2$ (Table 9, No. **65**) crystallizes in the orthorhombic space group Pbca − D$_{2h}^{15}$ (No. 61) with a = 26.514(18), b = 22.94(2), c = 22.346(10) Å, β = 90°; Z = 8, $D_c = 1.66$ g/cm^3. The structure is shown in **Fig. 59**. The molecule consists of an isosceles Os$_3$ triangle. All P atoms occupy equatorial coordination sites. In contrast to No. 58, the two P(C$_6$H$_5$)$_3$ ligands are bonded to different Os centers in a mutually trans configuration on Os(1) and Os(2), probably to minimize steric interactions [41].

References:

[1] Deeming, A. J.; Johnson, B. F. G.; Lewis, J. (J. Chem. Soc. A **1970** 897/901).

[2] Deeming, A. J.; Johnson, B. F. G.; Lewis, J. (J. Chem. Soc. A **1970** 2517/20).

[3] Crow, J. P.; Cullen, W. R. (Inorg. Chem. **10** [1971] 1529/31).

[4] Bradford, C. W.; Nyholm, R. S. (J. Chem. Soc. Dalton Trans. **1973** 529/33).

[5] Bhaduri, S.; Johnson, B. F. G.; Lewis, J.; Watson, D. J.; Zuccaro, C. (J. Chem. Soc. Chem. Commun. **1977** 477/8).

[6] Rivera, A. V.; Sheldrick, G. M. (Acta Crystallogr. B **34** [1978] 3372/4).

[7] Bhaduri, S.; Johnson, B. F. G.; Lewis, J.; Watson, D. J.; Zuccaro, C. (J. Chem. Soc. Dalton Trans. **1979** 557/61).

[8] Iwasaki, F.; Mays, M. J.; Raithby, P. R.; Taylor, P. L.; Wheatley, P. J. (J. Organomet. Chem. **213** [1981] 185/206).

[9] Deeming, A. J.; Peters, R. (J. Organomet. Chem. **235** [1982] 221/9).

[10] Deeming, A. J.; Peters, R.; Hursthouse, M. B.; Backer-Dirks, J. D. J. (J. Chem. Soc. Dalton Trans. **1982** 1205/11).

[11] Adams, R. D.; Horváth, I. T.; Segmüller, B. E.; Yang, L. W. (Organometallics **2** [1983] 144/8).

[12] Trukhacheva, V. A.; Burmakina, G. V.; Gubin, S. P.; Kovalev, Y. G.; Ioganson, A. A. (Izv. Akad. Nauk SSSR Ser. Khim. **1983** 2187/91; Bull. Acad. Sci. USSR Div. Chem. Sci. [Engl. Transl.] **32** [1983] 1973/6).

[13] Adams, R. D.; Horváth, I. T.; Segmüller, B. (J. Organomet. Chem **262** [1984] 243/52).

[14] Adams, R. D.; Hor, T. S. A.; Horvath, I. T. (Inorg. Chem. **23** [1984] 4733/8).

[15] Adams, R. D.; Horvath, I. T.; Mathur, P. (J. Am. Chem. Soc. **106** [1984] 6296/305).

[16] Wolf, M.; Lieto, J.; Matrana, B. A.; Arnold, D. B.; Knözinger, H. (J. Catal. **89** [1984] 100/10).

[17] Alex, R. F.; Pomeroy, R. K. (J. Organomet. Chem. **284** [1985] 379/84).

[18] Clucas, J. A.; Harding, M. M.; Smith, A. K. (J. Chem. Soc. Chem. Commun. **1985** 1280/1).

[19] Maksakov, V. A.; Ershova, V. A.; Bragina, I. V. (Izv. Akad. Nauk SSSR Ser. Khim. **1985** 2829/30; Bull. Acad. Sci. USSR Div. Chem. Sci. [Engl. Transl.] **34** [1985] 2627).

[20] Patel, V. D.; Cherkas, A. A.; Nucciarone, D.; Taylor, N. J.; Carty, A. J. (Organometallics **4** [1985] 1792/800).

[21] Adams, R. D.; Horváth, I. T.; Wang, S. (Inorg. Chem. **25** [1986] 1617/23).

[22] Bruce, M. I.; Horn, E.; Bin Shawkataly, O.; Snow, M. R.; Tiekink, E. R. T.; Williams, M. L. (J. Organomet. Chem. **316** [1986] 187/211).

[23] Cartwright, S.; Clucas, J. A.; Dawson, R. H.; Foster, D. F.; Harding, M. M.; Smith, A. K. (J. Organomet. Chem. **302** [1986] 403/12).

[24] Clucas, J. A.; Dawson, R. H.; Dolby, P. A.; Harding, M. M.; Pearson, K.; Smith, A. K. (J. Organomet. Chem. **311** [1986] 153/62).

[25] Sappa, E.; Nanni Marchino, M. L.; Predieri, G.; Tiripicchio, A.; Tiripicchio Camellini, M. (J. Organomet. Chem. **307** [1986] 97/105).

[26] Alex, R. F.; Pomeroy, R. K. (Organometallics **6** [1987] 2437/46).

[27] Alex, R. F.; Einstein, F. W. B.; Jones, R. H.; Pomeroy, R. K. (Inorg. Chem. **26** [1987] 3175/8).

[28] Ditzel, E. J.; Hanson, B. E.; Johnson, B. F. G.; Lewis, J. (J. Chem. Soc. Dalton Trans. **1987** 1285/8).

[29] Ditzel, E. J.; Johnson, B. F. G.; Lewis, J. (J. Chem. Soc. Dalton Trans. **1987** 1293/7).

[30] Ditzel, E. J.; Gómez-Sal, M. P.; Johnson, B. F. G.; Lewis, J.; Raithby, P. R. (J. Chem. Soc. Dalton Trans. **1987** 1623/30).

[31] Hodge, S. R.; Johnson, B. F. G.; Lewis, J.; Raithby, P. R. (J. Chem. Soc. Dalton Trans. **1987** 931/7).

[32] Süss-Fink, G.; Pellinghelli, M. A.; Tiripicchio, A. (J. Organomet. Chem. **320** [1987] 101/13).

[33] Bruce, M. I.; Liddell, M. J.; Hughes, C. A.; Patrick, J. M.; Skelton, B. W.; White, A. H. (J. Organomet. Chem. **347** [1988] 181/205).

[34] Deeming, A. J.; Kabir, S. E. (J. Organomet. Chem. **340** [1988] 359/66).

[35] Deeming, A. J.; Hardcastle, K. I.; Kabir, S. E. (J. Chem. Soc. Dalton Trans. **1988** 827/31).

[36] Colbran, S. B.; Irele, P. T.; Johnson, B. F. G.; Lahoz, F. J.; Lewis, J.; Raithby, P. R. (J. Chem. Soc. Dalton Trans. **1989** 2023/31).

[37] Arif, A. M.; Bright, T. A.; Heaton, D. E.; Jones, R. A.; Nunn, C. M. (Polyhedron **9** [1990] 1573/87).

[38] Field, J. S.; Haines, R. J.; Jay, J. A. (J. Organomet. Chem. **395** [1990] C 16/C 20).

[39] Ahmad, J. U.; Kabir, S. E.; Miah, M. A. (J. Bangladesh Chem. Soc. **4** [1991] 153/8).

[40] Azam, K. A.; Kabir, S. E.; Miah, A.; Day, M. W.; Hardcastle, K. I.; Rosenberg, E.; Deeming, A. J. (J. Organomet. Chem. **435** [1992] 157/67).

[41] Brown, M. P.; Dolby, P. A.; Harding, M. M.; Mathews, A. J.; Smith, A. K. (J. Chem. Soc. Dalton Trans. **1993** 1671/9).

[42] Dolby, P. A.; Harding, M. M.; Smith, A. K. (unpublished work from [41]).

[43] Kabir, S. E.; Miah, A.; Nesa, L.; Uddin, K.; Hardcastle, K. I.; Rosenberg, E.; Deeming, A. J. (J. Organomet. Chem. **492** [1995] 41/51).

Empirical Formula Index

In this index the compounds are listed in order of increasing carbon content. The empirical formulas are specified by linearized formulas. Ionic compounds are given in brackets; components of solvates and adducts are separated by a period.

Pages are printed in ordinary type, table numbers in bold face, and compound numbers in the tables in italics.

Ligand Formula Index

In this index all Os-bonded ligands are listed by their empirical formula (Hill formula with C and H first) in alphabetical order. The empirical ligand formulas are specified by their linearized formulas which distinct between structural isomers, but ligation mode is not taken into consideration. For each ligand the corresponding compounds are given. Compounds having more than one different ligand occur at more than one position. Only two ligands are not considered: H and CO.

Pages are printed in ordinary type, table numbers in bold face, and compound numbers in the tables in italics.

C_2HF_3N . . . . . $N=CH-CF_3$
    $(H)Os_3(CO)_{10}(N=CH-CF_3)$ 30/1, **2**, *1*

C_2HN_2 . . . . . . $N=CH-CN$
    $(H)Os_3(CO)_{10}(N=CH-CN)$ 31, **2**, *2*

$C_2H_2F_3N$ . . . . . $N-CH_2-CF_3$
    $(H)_4Os_3(CO)_8(N-CH_2-CF_3)$ 45, **3**, *2*
    $(H)_2Os_3(CO)_9(N-CH_2-CF_3)$ 46, **3**, *5*

$C_2H_3F_3N$ . . . . . $NH-CH_2-CF_3$
    $(H)Os_3(CO)_{10}(NH-CH_2-CF_3)$ 7, **1**, *6*

C_2H_3N . . . . . . $NC-CH_3$
    $(H)Os_3(CO)_9(NH_2)-NC-CH_3$ 4, **1**, *1*
    $(H)Os_3(CO)_9(NH-CH_3)-NC-CH_3$ 4/5, **1**, *2*

C_2H_4NO . . . . . $N(CH_3)CHO$
    $(H)Os_3(CO)_{10}[N(CH_3)CHO]$ 68/9, **5**, *7*

. . . . $NHC(CH_3)O$
    $(H)Os_3(CO)_{10}[NHC(CH_3)O]$ 67, **5**, *3*

C_2H_4NS . . . . . $SCHN(CH_3)$
    $(H)Os_3(CO)_9[SCHN(CH_3)]$ 94, **6**, *6*

$C_2H_5N_2$ . . . . . $N=C(CH_3)=NH_2$
    $(H)Os_3(CO)_9[N=C(CH_3)=NH_2]$ 101, **6**, *23*

. . . . $NHC(CH_3)NH$
    $(H)Os_3(CO)_{10}[NHC(CH_3)NH]$ 74, **5**, *21*
    $(H)Os_3(CO)_9[NH-C(CH_3)=NH]$ 101, **6**, *23*

C_2H_5S . . . . . . $S-C_2H_5$
    $(H)Os_3(CO)_8[(CH_3)_2P-CH_2-P(CH_3)_2](S-C_2H_5)$ 143, **9**, *47*
    $(H)Os_3(CO)_8[(C_6H_5)_2P-CH_2-P(C_6H_5)_2](S-C_2H_5)$ 144, **9**, *50*

$C_3H_3N_2$ . . . . . $1,2-N_2C_3H_3$
    $(H)Os_3(CO)_{10}(1,2-N_2C_3H_3)$ 76, **5**, *27*

$C_3H_4NS_2$ . . . . . $2-S-1,3-SNC_3H_4$
    $(H)Os_3(CO)_9(2-S-1,3-SNC_3H_4)$ 96/7, **6**, *12*

$C_3H_5N_2S$ . . . . . $2-S-1,3-N_2C_3H_5$
    $(H)Os_3(CO)_9(2-S-1,3-N_2C_3H_5)$ 97, **6**, *13*

C_3H_6N . . . . . . $N=C(CH_3)_2$
    $(H)Os_3(CO)_{10}[N=C(CH_3)_2]$ 32, **2**, *6*

. . . . $N=CH-C_2H_5$
    $(H)Os_3(CO)_{10}(N=CH-C_2H_5)$ 31, **2**, *3*

C_7H_6N . . . . . . N=CH–C_6H_5
   (H)Os$_3$(CO)$_{10}$(N=CH–C_6H_5) 31/2, **2**, *4*

C_7H_6NO . . . . . N(C_6H_5)CHO
   (H)Os$_3$(CO)$_{10}$[N(C_6H_5)CHO] 69, **5**, *10*

. . . . NHC(C_6H_5)O
   (H)Os$_3$(CO)$_{10}$[NHC(C_6H_5)O] 68, **5**, *6*

C_7H_6NS . . . . . SCHN(C_6H_5)
   (H)Os$_3$(CO)$_{10}$[N(C_6H_5)CHS] 72, **5**, *16*
   (H)Os$_3$(CO)$_9$[SCHN(C_6H_5)] 94/5, **6**, *7*

$C_7H_6N_3$ . . . . . 2-NH-1,3-$N_2C_7H_5$
   (H)Os$_3$(CO)$_{10}$(2-NH-1,3-$N_2C_7H_5$) 75, **5**, *24*

C_7H_7N . . . . . . N–CH$_2$–C_6H_5
   (H)$_2$Os$_3$(CO)$_9$(N–CH$_2$–C_6H_5) 46, **3**, *7*

. . . . N–C_6H_4-4-CH$_3$
   (H)$_2$Os$_3$(CO)$_9$(N–C_6H_4-4-CH$_3$) 47, **3**, *11*
   [(C_6H_5)$_3$P=N=P(C_6H_5)$_3$][Os$_3$(CO)$_9$(N–C_6H_4-4-CH$_3$)(Cl)] 49, **3**, *16*
   [(C_6H_5)$_3$P=N=P(C_6H_5)$_3$][Os$_3$(CO)$_9$(N–C_6H_4-4-CH$_3$)(Br)] 50, **3**, *18*
   [(C_6H_5)$_3$P=N=P(C_6H_5)$_3$][Os$_3$(CO)$_9$(N–C_6H_4-4-CH$_3$)(I)] 50, **3**, *20*
   [(C_6H_5)$_3$P=N=P(C_6H_5)$_3$][Os$_3$(CO)$_9$(N–C_6H_4-4-CH$_3$)(Br)] 50/1, **3**, *22*
   [(C_6H_5)$_3$P=N=P(C_6H_5)$_3$][Os$_3$(CO)$_9$(N–C_6H_4-4-CH$_3$)(I)] 51, **3**, *24*
   Os$_3$(CO)$_{10}$(N–C_6H_4-4-CH$_3$) 53, **3**, *28*

$C_7H_7N_2$ . . . . . N=N(C_6H_4-4-CH$_3$)
   (H)Os$_3$(CO)$_{10}$(N=N–C_6H_4-4-CH$_3$) 35/6, **2**, *14*
   (H)Os$_3$(CO)$_{10}$[N=N(C_6H_4-4-CH$_3$)] 78, **5**, *32*

. . . . NH–N=CH–C_6H_5
   (H)Os$_3$(CO)$_{10}$(NH–N=CH–C_6H_5) 14, **1**, *25*

$C_7H_7N_2S$ . . . . . SC(NH$_2$)N(C_6H_5)
   (H)Os$_3$(CO)$_9$[SC(NH$_2$)N(C_6H_5)] 95/6, **6**, *10*

C_7H_7S . . . . . . S–C_6H_4-4-CH$_3$
   (H)Os$_3$(CO)$_8$[(C_6H_5)$_2$P–CH$_2$–P(C_6H_5)$_2$](S–C_6H_4-4-CH$_3$) 146, **9**, *56*
   [(H)$_2$Os$_3$(CO)$_8$((C_6H_5)$_2$P–CH$_2$–P(C_6H_5)$_2$)(S–C_6H_4-4-CH$_3$)][PF$_6$] . . 146, **9**, *57*

C_7H_8N . . . . . . NH–CH$_2$–C_6H_5
   (H)Os$_3$(CO)$_{10}$(NH–CH$_2$–C_6H_5) 9/10, **1**, *14*

. . . . NH–C_6H_4-4-CH$_3$
   (H)Os$_3$(CO)$_{10}$(NH–C_6H_4-4-CH$_3$) 12, **1**, *20*

$C_7H_8NO_2S$ . . . . NH–S(=O)$_2$–C_6H_4-4-CH$_3$
   (H)Os$_3$(CO)$_{10}$[NH–S(=O)$_2$–C_6H_4-4-CH$_3$] 13, **1**, *23*

$C_7H_8N_2$ . . . . . $N=NH(C_6H_4-4-CH_3)$
   $[(H)Os_3(CO)_{10}(N=NH(C_6H_4-4-CH_3))][BF_4]$ 78/9, **5**, *33*
   $[(H)Os_3(CO)_{10}(N=NH(C_6H_4-4-CH_3))]Cl$ 87

$C_7H_8N_3$ . . . . . $NH-N=N(CH_2-C_6H_5)$
   $(H)Os_3(CO)_{10}[NH-N=N(CH_2-C_6H_5)]$ 80, **5**, *37*

$C_7H_{15}N_2$ . . . . . $N(C_3H_7-i)CHN(C_3H_7-i)$
   $(H)Os_3(CO)_9[N(C_3H_7-i)CHN(C_3H_7-i)]$ 102, **6**, *24*
   $[(H)_2Os_3(CO)_9(N(C_3H_7-i)CHN(C_3H_7-i))][O_2C-CF_3]$ 102/3, **6**, *25*

. . . . $N(C_3H_7-i)CHN-C_3H_7-i$
   $(H)Os_3(CO)_8[N(C_3H_7-i)-CH=N-C_3H_7-i]-P(CH_3)_2-C_6H_5$ 136, **9**, *22*

C_8H_8N . . . . . . $N=C(CH_3)-C_6H_5$
   $(H)Os_3(CO)_{10}[N=C(CH_3)-C_6H_5]$ 33, **2**, *8*

C_8H_8NO . . . . . $N(C_6H_4-4-CH_3)CHO$
   $(H)Os_3(CO)_{10}[N(C_6H_4-4-CH_3)CHO]$ 69, **5**, *9*

C_8H_8NS . . . . . $SCHN(C_6H_4-4-CH_3)$
   $(H)Os_3(CO)_9[SCHN(C_6H_4-4-CH_3)]$ 95, **6**, *9*

$C_8H_8N_3$ . . . . . $NH-N=N(C(=CH_2)-C_6H_5)$
   $Os_3(CO)_{10}[NH-N=N(C(=CH_2)-C_6H_5)]$ 80, **5**, *38*

$C_8H_9N_2$ . . . . . $NH-N=C(CH_3)-C_6H_5$
   $(H)Os_3(CO)_{10}[NH-N=C(CH_3)-C_6H_5]$ 14, **1**, *27*

. . . . $2-(NC_4H_6-2)-NC_4H_3$
   $(H)Os_3(CO)_9[2-(NC_4H_6-2)-NC_4H_3]$ 101, **6**, *22*

$C_8H_{11}P$ . . . . . $P(CH_3)_2-C_6H_5$
   $Os_3S_2(CO)_7[P(CH_3)_2-C_6H_5]_2$ 121, **8**, *4*
   $Os_3(CO)_8[P(CH_3)_2-C_6H_5]_4$ 127
   $Os_3S_2(CO)_8-P(CH_3)_2-C_6H_5$ 133/4, **9**, *17*
   $(H)Os_3(CO)_8[N(C_3H_7-i)-CH=N-C_3H_7-i]-P(CH_3)_2-C_6H_5$ 136, **9**, *22*
   $(H)Os_3(CO)_8(2-NH-NC_5H_4)-P(CH_3)_2-C_6H_5$ 136, **9**, *23*

$C_8H_{12}N_2$ . . . . . $c-C_3H_5-N=CH-CH=N-C_3H_5-c$
   $Os_3(CO)_{10}(c-C_3H_5-N=CH-CH=N-C_3H_5-c)$ 59

$C_8H_{16}F_3NP$ . . . . $N-CH(CF_3)-P(C_2H_5)_3$
   $(H)Os_3(CO)_{10}[N-CH(CF_3)-P(C_2H_5)_3]$ 36, **2**, *15*

$C_8H_{16}N_2$ . . . . . $i-C_3H_7-N=CH-CH=N-C_3H_7-i$
   $Os_3(CO)_{10}(i-C_3H_7-N=CH-CH=N-C_3H_7-i)$ 61/2, **4**, *5*

$C_8H_{18}P$ . . . . . $P(C_4H_9-t)_2$
   $(H)_2Os_3(CO)_6[P(C_4H_9-t)_2]_2-PH(C_4H_9-t)_2$ 117, **7**, *3*
   $(H)_2Os_3(CO)_8[P(C_4H_9-t)_2]_2$ 136/7, **9**, *24*

Transition Metal Cross Reference

All organoosmium compounds containing additional transition metals are listed in alphabetical order of these metals. Compounds with more than one additional metal are listed under each of them.

Pages are printed in ordinary type, table numbers in bold face, and compound numbers in the tables in italics.

Physical Constants and Conversion Factors

Avogadro constant N_A(or L) = 6.02214 × 10²³ mol⁻¹

Faraday constant F = 9.64853 × 10⁴ C/mol

molar gas constant R = 8.31451 J·mol⁻¹·K⁻¹

molar volume (ideal gas) V_m = 2.24141 × 10¹ L/mol
(273.15 K, 101325 Pa)

Planck constant h = 6.62608 × 10⁻³⁴ J·s

elementary charge e = 1.60218 × 10⁻¹⁹ C

electron mass m_e ≃ 9.10939 × 10⁻³¹ kg

proton mass m_p = 1.67262 × 10⁻²⁷ kg

1 kg = 2.205 pounds

1 m = 3.937 × 10¹ inches = 3.281 feet

1 m³ = 2.642 × 10² gallons (U.S.)

1 m³ = 2.200 × 10² gallons (Imperial)

Force	N	dyn	kp
1 N	1	10⁵	1.019716 × 10⁻¹
1 dyn	10⁻⁵	1	1.019716 × 10⁻⁶
1 kp	9.80665	9.80665 × 10⁵	1

Pressure	Pa	bar	kp/m²	at	atm	Torr	lb/in²
1 Pa = 1N/m²	1	10⁻⁵	1.019716 × 10⁻¹	1.019716 × 10⁻⁵	9.86923 × 10⁻⁶	7.50062 × 10⁻³	1.450378 × 10⁻⁴
1 bar = 10⁶ dyn/cm²	10⁵	1	1.019716 × 10⁴	1.019716	9.86923 × 10⁻¹	7.50062 × 10²	1.450378 × 10¹
1 kp/m² = 1 mm H₂O	9.80665	9.80665 × 10⁻⁵	1	10⁻⁴	9.67841 × 10⁻⁵	7.35559 × 10⁻²	1.422335 × 10⁻³
1 at (technical)	9.80665 × 10⁴	9.80665 × 10⁻¹	10⁴	1	9.67841 × 10⁻¹	7.35559 × 10²	1.422335 × 10¹
1 atm = 760 Torr	1.01325 × 10⁵	1.01325	1.033227 × 10⁴	1.033227	1	7.60 × 10²	1.469595 × 10¹
1 Torr = 1 mm Hg	1.333224 × 10²	1.333224 × 10⁻³	1.359510 × 10¹	1.359510 × 10⁻³	1.315789 × 10⁻³	1	1.933678 × 10⁻²
1 lb/in² = 1 psi	6.89476 × 10³	6.89476 × 10⁻²	7.03069 × 10²	7.03069 × 10⁻²	6.80460 × 10⁻²	5.17149 × 10¹	1

Work, Energy, Heat

Work, Energy, Heat	J	kW·h	kcal	Btu	eV
1 J = 1 W·s = 1 N·m = 10^7 erg	1	2.778×10^{-7}	2.39006×10^{-4}	9.4781×10^{-4}	6.242×10^{18}
1 kW·h	3.6×10^6	1	8.604×10^2	3.41214×10^3	2.247×10^{25}
1 kcal	4.1840×10^3	1.1622×10^{-3}	1.	3.96566	2.6117×10^{22}
1 Btu (British thermal unit)	1.05506×10^3	2.93071×10^{-4}	2.5164×10^{-1}	1	6.5858×10^{21}
1 eV	1.602×10^{-19}	4.450×10^{-26}	3.8289×10^{-23}	1.51840×10^{-22}	1

$1\ cm^{-1} \cong 1.239842 \times 10^{-4}\ eV$ $1\ Hz \cong 4.135669 \times 10^{-15}\ eV$

$2\ \text{Rydberg (Ry)} = 1\ \text{hartree} = 27.2114\ eV$ $1\ eV \cong 23.0578\ \text{kcal/mol}$

Power

Power	kW	hp	kp·m·s^{-1}	kcal/s
1 kW = 10^3 J/s	1	1.35962	1.01972×10^2	2.39006×10^{-1}
1 hp (horsepower, metric)	7.3550×10^{-1}	1	7.5×10^1	1.7579×10^{-1}
1 kp·m·s^{-1}	9.80665×10^{-3}	1.333×10^{-2}	1	2.34384×10^{-3}
1 kcal/s	4.1840	5.6886	4.26650×10^2	1

References:

Mills, I. (Ed.), International Union of Pure and Applied Chemistry, Quantities, Units and Symbols in Physical Chemistry, Blackwell Scientific Publications, Oxford 1988.

The International System of Units (SI), National Bureau of Standards Spec. Publ. 330 [1972].

Landolt-Börnstein, 6th Ed., Vol. II, Pt. 1, 1971, pp. 1/14.

ISO Standards Handbook 2, Units of Measurement, 2nd Ed., Geneva 1982.

Cohen, E. R., Taylor, B. N., Codata Bulletin No. 63, Pergamon, Oxford 1986.

Key to the Gmelin System
of Elements and Compounds

System Number	Symbol	Element
1		Noble Gases
2	H	Hydrogen
3	O	Oxygen
4	N	Nitrogen
5	F	Fluorine
6	**Cl**	**Chlorine**
7	Br	Bromine
8	I	Iodine
8a	At	Astatine
9	S	Sulfur
10	Se	Selenium
11	Te	Tellurium
12	Po	Polonium
13	B	Boron
14	C	Carbon
15	Si	Silicon
16	P	Phosphorus
17	As	Arsenic
18	Sb	Antimony
19	Bi	Bismuth
20	Li	Lithium
21	Na	Sodium
22	K	Potassium
23	NH_4	Ammonium
24	Rb	Rubidium
25	Cs	Caesium
25a	Fr	Francium
26	Be	Beryllium
27	Mg	Magnesium
28	Ca	Calcium
29	Sr	Strontium
30	Ba	Barium
31	Ra	Radium
32	**Zn**	**Zinc**
33	Cd	Cadmium
34	Hg	Mercury
35	Al	Aluminium
36	Ga	Gallium

System Number	Symbol	Element
37	In	Indium
38	Tl	Thallium
39	Sc, Y La—Lu	Rare Earth Elements
40	Ac	Actinium
41	Ti	Titanium
42	Zr	Zirconium
43	Hf	Hafnium
44	Th	Thorium
45	Ge	Germanium
46	Sn	Tin
47	Pb	Lead
48	V	Vanadium
49	Nb	Niobium
50	Ta	Tantalum
51	Pa	Protactinium
52	**Cr**	**Chromium**
53	Mo	Molybdenum
54	W	Tungsten
55	U	Uranium
56	Mn	Manganese
57	Ni	Nickel
58	Co	Cobalt
59	Fe	Iron
60	Cu	Copper
61	Ag	Silver
62	Au	Gold
63	Ru	Ruthenium
64	Rh	Rhodium
65	Pd	Palladium
66	Os	Osmium
67	Ir	Iridium
68	Pt	Platinum
69	Tc	Technetium[1]
70	Re	Rhenium
71	Np,Pu...	Transuranium Elements

HCl

$CrCl_2$

$ZnCrO_4$

$ZnCl_2$

Material presented under each Gmelin System Number includes all information concerning the element(s) listed for that number plus the compounds with elements of lower System Number.

For example, zinc (System Number 32) as well as all zinc compounds with elements numbered from 1 to 31 are classified under number 32.

[1] A Gmelin volume titled "Masurium" was published with this System Number in 1941.

A Periodic Table of the Elements with the Gmelin System Numbers is given on the Inside Front Cover